SpringerBriefs in Latin American Studies

Series Editors

Jorge Rabassa, Lab Geomorfología y Cuaternario, CADIC-CONICET, Ushuaia, Tierra del Fuego, Argentina

Eustógio Wanderley Correia Dantas, Departamento de Geografia, Centro de Ciências, Universidade Federal do Ceará, Fortaleza, Ceará, Brazil

SpringerBriefs in Latin American Studies promotes quality scientific research focusing on Latin American countries. The series accepts disciplinary and interdisciplinary titles related to geographical, environmental, cultural, economic, political and urban research dedicated to Latin America. The Series will publish compact volumes (50 to 125 pages) refereed by a region or country expert specialized in Latin American studies. We offer fast publication time of 8 to 12 weeks after acceptance. The series aims to raise the profile of Latin American studies, showcasing important works developed focusing on the region. It is aimed at researchers, students, and everyone interested in Latin American topics.

Flávio Rodrigues do Nascimento

Global Environmental Changes, Desertification and Sustainability

Springer

Flávio Rodrigues do Nascimento
Federal University of Ceará
Fortaleza, Ceará, Brazil

ISSN 2366-763X   ISSN 2366-7648 (electronic)
SpringerBriefs in Latin American Studies
ISBN 978-3-031-32949-4   ISBN 978-3-031-32947-0 (eBook)
https://doi.org/10.1007/978-3-031-32947-0

This Springer imprint is published by the registered company Springer Nature Switzerland AG
The registered company address is: Gewerbestrasse 11, 6330 Cham, Switzerland

Introduction

Human activities recorded in the Holocene, without scales and precedent, have been causing profound impacts on the planet, affecting air, water, and soil resources. Moreover, since the middle of the last century, environmental degradation has been increasing with the "Great Acceleration" of consumption and productive activity.

Global environmental problems gain momentum, surpassing "Planetary Boundaries", indicating the entry into the Anthropocene by human interference in the Biosphere and triggering the "Sixth Great Extinction". The climate crisis stands out as the cause and effect of these problems, with the advance of extreme events—especially droughts and desertification.

The speed of environmental changes, the increase in the concentration of carbon dioxide (CO_2) in the atmosphere and its abnormal warming, the greenhouse effect, and the increase in desertified areas are linked to climate change. Trend tables indicate that global changes are worsened by changes in temperature and rainfall resulting from climate change, with strong anthropogenetic influence.

The drylands and the poorest countries would be the most affected by these transformations, which involve aggravation of ecological, social, and economic problems affecting sustainability. Desertification is widespread as a result of this. In the effects of global and climate change, desertification is a complex and harmful phenomenon globally. It involves several factors and causes affecting natural, rural, and urban areas, imposing enormous challenges for governments, civil society, and the private sector, now and intertemporally. The Sustainable Development Goals (SDGs), especially Goal 13 (urgent measures to combat climate change and its impacts), try to mitigate these issues.

However, difficulties persist in considering a consensual definition of desertification until today, even with the climate crisis and the increase in droughts in the world. This represents an obstacle to confronting the problem on a global level. These aspects denote the dimension of the problem's complexity.

The causes and repercussions of desertification in the world are varied in type or origin, degree, and scope. As for the origin, desertification can be caused by physical and human factors. Its magnitude corresponds to the degree of intensity of the phenomenon and its related effects. Comprehensiveness has to do with its scale and scope. The leading causes of desertification are associated with degrading traditional agriculture, irrigated and poorly managed agriculture, soil salinization, contingencies of climate change, and extreme events.

However, structural factors, such as income concentration and inadequacies of some economic, cultural, and technological activities to environmental conditions, hinder desertification treatment. Furthermore, more than climatic contingencies, human interference in the biophysical environment causes degradation and desertification.

This demands regional and local studies to understand better the environmental problems of areas affected by this problem. Recognition of the extent of the drylands of the globe and their geographic distribution reveal greater accuracy in diagnosing and monitoring the problem. Specifically, the management of hydrographic basins and the water resources must consider the environmental susceptibility and the risks of salinization of soils worldwide, especially in drylands and irrigated perimeters. Soil salinization and sodicity are among the main problems of environmental degradation and can cause or increase desertification.

Therefore, the management of natural resources, environmental management, monitoring of evolutionary scenarios, and monitoring and mitigation of the climate crisis, especially in the drylands, must follow ecological, economic, and social sustainability precepts. Different nations need to establish a new global environmental order. Their priority agenda should highlight global environmental changes, the climate crisis, and desertification.

Contents

Chapter 1
Society × Nature—Anthropocene and Limits of Balance on Earth

Abstract Today's global transformations have complex roots linked to geological time and the ecological and environmental dynamics of the Planet, emphasizing the occurrences recorded in the Holocene. The Industrial Revolution and other technical-scientific revolutions affected and accelerated the relationship between society and nature. The consumption increased, and the production of space transformed continental areas, impacting the air and water. Some works claim that the second half of the twentieth century displayed the "Great Acceleration", a critical time within the Anthropocene following the Holocene. Therefore, global environmental problems gain momentum, exceeding "Planetary Boundaries", indicating the entry into the "Sixth Great Extinction". In this situation, the climate crisis stands out as cause and effect, with the advance of extreme events becoming frequent, producing, for example, droughts and an increase in desertification.

Keywords Society × Nature · Holocene/anthropocene · Global environmental changes · Climate crisis limits · Planetaries

1.1 Global Environmental Changes and Threatened Sustainability

Climatic conditions with unparalleled stability predominated in the planet's environmental conditions in the last 100 thousand years. This stability occurred with the entry into the Holocene (current time of the Quaternary period of the Cenozoic era), about 11,000 years before the present/AP, after the last glacial period. This fact allowed the global population of 5 million people to develop and expand the activities of plantations, animal domestication, and the construction of population clusters, which then became urban centers. In addition, the Holocene defined a critical increase in migrations worldwide.

With the accumulation of time and the spatial actions mediated by the achievements of humanity, the current geological era has our society primarily responsible for changes on the planet (Rockström et al. 2009; SRC 2022a). With the triggering

© The Author(s), under exclusive license to Springer Nature Switzerland AG 2023
F. Rodrigues do Nascimento, *Global Environmental Changes, Desertification and Sustainability*, SpringerBriefs in Latin American Studies, https://doi.org/10.1007/978-3-031-32947-0_1

of various contingencies, we are witnessing deregulation of the Biosphere; what would be the entry into a new geological era called the Anthropocene by Mathis Wackernagel and William Rees in the early 1990s (UNESCO 2002).

Despite this, Steffen et al., from the Stockholm Resilience Center (SRC), in a paper published in Science in 2015 (Planetary boundaries: Guiding human development on a changing planet), state that the activities of society have driven climate change, the loss of biodiversity, changes in nutrient cycles (Nitrogen/N and Phosphorus/P), and land use across borders into unprecedented territory. In addition, humanity affected the natural resources of water, air, and soil, impacting essential cycles in planning (biogeochemicals—water, carbon, oxygen, nitrogen) thus putting pressure on the conditions for sustaining life.

It seems like we have already entered the "Sixth Extinction" (Kolbert 2014): 40% of the world's mammals have already lost 80% of their habitats between 1900 and 2015. The modification of ecosystems, under the combined effects of mass consumption, expressive population increase with deep social inequalities, urban growth, triggered the phenomenon called "the great acceleration" (UNESCO 2002).

In this scenario, SRC pointed out the risk of breaking the earth's natural balance and resilience, defining nine limits or interconnected parameters essential to maintain the planet's stability. Specific measures have been defined for each of these planetary boundaries. As well as an area of safe operation so that such limits are not exceeded and another area showing the risk of crossing these borders—whose area is in increasing danger. Once it does not cross such borders, humanity will be able to prosper for generations. Unfortunately, of the nine Planetary Boundaries/LP, the first four listed below have already been crossed; there are still three within the safe zone, and two are a big unknown as there are no parameters to monitor them: the chemical pollution and atmospheric aerosol loading (Rockström et al. 2009a, b; Steffen et al. 2015; SRC 2022a).

This approach to global sustainability defines LPs within which humanity can operate safely. Transgressing any planetary boundaries can be deleterious or catastrophic, given the risk of crossing thresholds that would trigger non-linear and abrupt environmental changes within continental to planetary-scale systems. The LPs are Climate Change; Biosphere Integrity; Changes in land use; Biogeochemical Flows; Depletion of Atmospheric Ozone; Use of fresh water; Ocean Acidification; Loading of atmospheric aerosols; e., Incorporation of New Entities (Rockström et al. 2009).

The seven parameterized limits are:

These seven are climate change (CO_2 concentration in the atmosphere <350 ppm and/or a maximum change of +1 W m^{-2} in radiative forcing); ocean acidification (mean surface seawater saturation state with respect to aragonite $\geq$ 80% of pre-industrial levels); stratospheric ozone (<5% reduction in O_3 concentration from pre-industrial level of 290 Dobson Units); biogeochemical nitrogen (N) cycle (limit industrial and agricultural fixation of N_2 to 35 Tg N yr^{-1}) and phosphorus (P) cycle (annual P inflow to oceans not to exceed 10 times the natural background weathering of P); global freshwater use (<4000 km^3 yr^{-1} of consumptive use of runoff resources); land system change (<15% of the ice-free land surface under cropland); and the rate at which biological diversity is lost (annual rate of <10 extinctions per million species). (Rockström et al. 2009, p. 1)

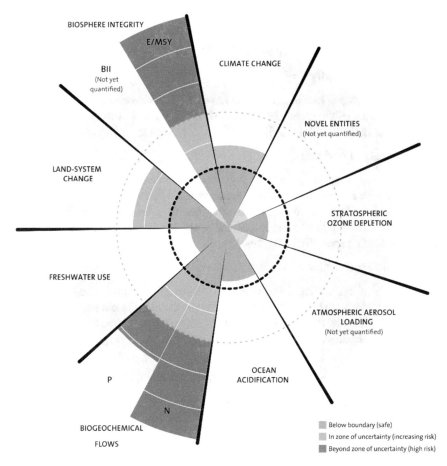

Fig. 1.1 Current status of the control variables for seven of the planetary boundaries. *Source* J. Lokrantz/Azote based on Steffen et al. (2015)

Figure 1.1 shows the status of the control variables for the seven measured LPs.

1.2 Planetary Limits/LP: Recompilation

Without exhausting the Planetary Boundaries representation, and in terms of something straightforward for its complex discussion, each one of them was synthesized:

1st Climate Change

Compared to the pre-industrial period, the rise in planetary temperature by 1.1 °C compared to the pre-industrial period has caused global environmental changes.

Compared to just over 50 years ago, there are five times more meteorological disasters, with costs seven times higher with devastation and death. Climate change is one of the central limits of its global influence. This work will later deal with this issue.

2nd Biosphere Integrity

There is an increasing loss of Biodiversity and species extension, with about 12.5% of lost life forms (Sixth Great Extinction). Unlike the 1st LP, this limit has already exceeded the risk zone for high risk. Ecosystem services are being lost.

3rd Changes in Land Use

The production of space worldwide is transforming forests, grasslands, swamps, tundras, and other phytoecological systems, creating space for agriculture. Deforestation strongly impacts climate regulation, causing a decline in biodiversity and food production. In addition, improper soil management and salinization increase desertified lands, especially in Drylands.

4th Biogeochemical Flows

Biogeochemical cycles maintain the dynamics and life on earth. They are greatly affected, especially by N and P. Although they are essential to plants, the excessive use of fertilizers affects soils and organisms. Besides, fertilizers are being drained by rivers and their hydrographic basins going to seas and oceans, surpassing the marine ecological limits.

5th Destruction of Ozone/Stratospheric O_3

One of the limits maintained by human actions, Chlorofluorocarbons (CFCs), have been banned since the Montreal Protocol. The atmospheric O_3 has been recovering since then, reducing the incidence of cancers and various environmental damage, especially in the thermal balance of the planet.

6th Use of Fresh Water

Although still in the safe area, it is close to the risk area. Water security seems to have been broken, especially in the drylands and desertification areas.

7th Ocean Acidification

Also in the security area, but close to the risk area. It has an extra layer of risk because Ocean Acidification triggered several mass extinctions in Geological History.

8th Loading of Atmospheric Aerosols

This issue has no baseline for parameterization, as it is a new component arising from the use of aerosols by humans—microscopic particles caused by fires and fossil fuels. It produces human deaths by inhalation of contaminated air and losses of other living organisms and alters the tropical monsoon system.

9th Incorporation of New Entities

The new entities are elements or organisms that human actions and new substances have modified. There are hundreds of thousands of entities ranging from radioactive materials to microplastics.

Figure 1.2 organizes a table with the LP, systemic processes, parameters involved, proposed limits, current situation of values, and pre-industrial data for comparison.

So, it is argued that the original organization of the Biosphere has been disrupted. Furthermore, the interactions between the Biosphere Integrity LP and other planetary boundaries negatively impact and produce complex effects generating uncertainties (Fig. 1.3).

PLANETARY BOUNDARIES

Earth-system process	Parameters	Proposed boundary	Current status	Pre-industrial value
Climate change	(i) Atmospheric carbon dioxide concentration (parts per million by volume)	350	387	280
	(ii) Change in radiative forcing (watts per metre squared)	1	1.5	0
Rate of biodiversity loss	Extinction rate (number of species per million species per year)	10	>100	0.1–1
Nitrogen cycle (part of a boundary with the phosphorus cycle)	Amount of N_2 removed from the atmosphere for human use (millions of tonnes per year)	35	121	0
Phosphorus cycle (part of a boundary with the nitrogen cycle)	Quantity of P flowing into the oceans (millions of tonnes per year)	11	8.5–9.5	~1
Stratospheric ozone depletion	Concentration of ozone (Dobson unit)	276	283	290
Ocean acidification	Global mean saturation state of aragonite in surface sea water	2.75	2.90	3.44
Global freshwater use	Consumption of freshwater by humans (km^3 per year)	4,000	2,600	415
Change in land use	Percentage of global land cover converted to cropland	15	11.7	Low
Atmospheric aerosol loading	Overall particulate concentration in the atmosphere, on a regional basis	To be determined		
Chemical pollution	For example, amount emitted to, or concentration of persistent organic pollutants, plastics, endocrine disrupters, heavy metals and nuclear waste in, the global environment, or the effects on ecosystem and functioning of Earth system thereof	To be determined		

Fig. 1.2 Planetary boundaries and global sustainability. *Source* Steffen (2015) and SRC (2022b)

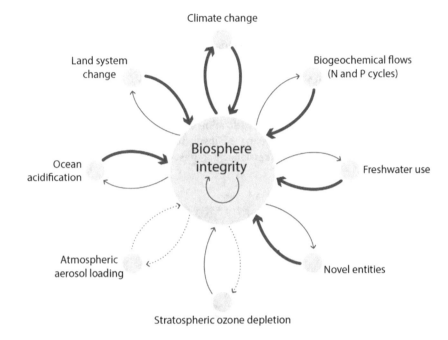

Fig. 1.3 The interaction between the planetary boundaries of biosphere integrity and other planetary boundaries. Adapted from: Mace et al. (2014) in Steffen (2015)

In this uproar, Global Environmental Changes take place, highlighting the Planet's Climate Crisis with its extreme events. Among these are the droughts. Moreover, there is Desertification as a dialectical cause and consequence of droughts and droughts.

References

Kolbert E (2014) A sexta extinção—uma história não natural. Ed. Intrinseca, Rio de Janeiro, 284p
Mace GM et al (2014) Approaches to defining a planetary boundary for biodiversity. Glob Environ Change 28:289–297
Rockström J, Steffen W, Noone K et al (2009a) A safe operating space for humanity. Nature 461:472–475. https://doi.org/10.1038/461472a
Rockström JW et al (2009b) Planetary boundaries: exploring the safe operating space for humanity. Ecol Soc 14(2):32 [Online]. http://www.ecologyandsociety.org/vol14/iss2/art32/
Steffen W et al (2015) Planetary boundaries: guiding human development on a changing planet. Science 347:11. https://doi.org/10.1126/science.1259855

Stockholm Resilience Centre (SRC) (2022a) Planetary boundaries. Cited on https://www.stockh olmresilience.org/research/planetary-boundaries.html, 17 Feb 2022

Stockholm Resilience Centre (SRC) (2022b) Figure and data for updated planetary boundaries. Cited on https://www.stockholmresilience.org/research/planetary-boundaries/planetary-bounda ries-data.html, 18 Feb 2022

UNESCO (2022) Um glossário para o antropoceno. Cited on pt.unesco.org, 17 Feb 2022

Chapter 2
Global Environmental Change, Climate Crisis and Desertification

Abstract The speed of environmental changes, the increase in the carbon dioxide (CO_2) concentration in the atmosphere and its abnormal warming, the greenhouse effect, and the increase in desertified areas are linked to climate change. Trend tables indicate that global changes are worsened by changes in temperature and rainfall resulting from climate change, with a strong anthropogenetic influence. Reports from the United Nations/UN, the IPCC (Intergovernmental Panel on Climate Change), and the World Meteorological Organization (WMO) indicate that the world has already suffered a 1.0 °C global warming above the pre-industrial levels, with a variation between 0.8 and 1.2 °C. Drylands and the poorest countries would be the most affected by these transformations, which involve aggravation of ecological, social, and economic problems. The desertification would spread due to this situation. Among the effects of global climate change, desertification is one of the most complex and harmful. It involves several factors and causes affecting natural, rural, and urban areas, posing major challenges for governments, civil society, the private sector, and future generations. The discussion on the Sustainable Development Goals (SDGs), especially Goal 13 (urgent measures to combat climate change and its impacts), tries to mitigate these issues. However, the mechanisms linked to climate change are still elements of skepticism for scientific denialists.

Keywords Global change · Climate change · Environmental degradation · Desertification

2.1 Climate Change and International Agreements

Worldwide climate change is one of the most discussed environmental problems today (UN 2020). Environmental problems have worsened with the increased complexity of human impacts on the environment. The cumulative importance of these impacts was highlighted, provoking a discussion about global climate change and its effects, among them desertification.

© The Author(s), under exclusive license to Springer Nature Switzerland AG 2023
F. Rodrigues do Nascimento, *Global Environmental Changes, Desertification and Sustainability*, SpringerBriefs in Latin American Studies, https://doi.org/10.1007/978-3-031-32947-0_2

Changes that would take millions of years can now occur in a few years or decades. The increase in the carbon dioxide (CO_2) concentration in the atmosphere and its abnormal heating, triggering the greenhouse effect, and the increase in desert areas are among the main consequences. As for the emission of greenhouse gases (GHG, Table 2.1), many mechanisms are linked to climate change: thermal pollution, changes in albedo, the extension of irrigated land, alteration of marine currents, and diversion of fresh water to the ocean.

Historically, the wealthiest countries have been and still are mainly responsible for these pollutant gas emissions: the United States of America (USA), Japan, and Germany. China can be included in the current century, already occupying second place in this regard! Meanwhile, despite being poorer and the late industrialization that took place after World War II, the newly industrialized countries, such as the BRICS countries, surpass the rich nations in one item in this type of pollution, which is the emission of GGE resulting from the burning of biomass. In addition to Brazil, Mexico, and India, China appears again among countries that are owners of significant forest reserves and members of the restricted group of megadiverse countries.

The Intergovernmental Panel on Climate Change (IPCC) conference, held in Paris on 02/04/2008, ratified the problem of desertification as a possible consequence of global warming. The conference pointed to reforestation as a sink and sequester of carbon for desertification treatment and better performance of activities related to global warming. Excluding the industrial treatment of CO_2, which technology has not yet wholly solved, reforestation is one of the only means of reducing the amount of carbon produced by combustion and an alternative for a possible alleviation of adverse climatic effects (http://www.ipcc.ch, 2021).

It is clear that there is a long cycle of temperature variations on the planet, and warming is natural and inevitable. However, in recent decades, warming has reached an intensity never before recorded, leading to the belief that anthropogenetic GHG emissions potentiate this phenomenon, maximizing the influence of the greenhouse effect on climate change and, consequently, on desertification.

Table 2.1 Origin of the main greenhouse gases/GHG

Gas	Natural sources	Sources derived from human activities
Carbon dioxide	Terrestrial biosphere; oceans	Uses of fossil fuels; cement production; land-use change
Methane	Natural wetlands; termites (termites); oceans and freshwater collections	Fossil fuels; fermentation of animal waste; rice paddies; biomass burning; landfills; domestic sewage
Nitrous oxide	Oceans; soils of tropical and temperate regions	Nitrogen-based fertilizers; industry; changes in land use; intensive livestock
Chlorofluorocarbon (CFC)	There is not	Rigid and flexible foams; aerosol propellants; Teflon polymers; industrial solvents

Source Bensunsan (2002) and do Nascimento (2016)

The understanding that among the leading global changes is climate change by induction or acceleration of human activities prevails. According to Filho et al. (1994), one of the first IPCC reports points out that, by the year 2030, the CO_2 concentration would double the current one. Furthermore, the 2001 IPCC report (third document) confirms that there is an elevation in the average temperature of the planet as a result of human activities, with increased temperatures that would fluctuate between 1 and 3.5 °C average for the next 100 years due to the duplication of the CO_2 concentration compared to the pre-industrial phase (Furriela 2002).

According to Furriela (2002), Gomes and Zanella (2021), and IPCC (2021), there is evidence that proves climate change in the last century, which would reinforce the need to develop the following principles: increase in the concentration of gases such as carbon dioxide (CO_2), methane (CH_4) and nitrous oxide (N_2O), which, between 1750 and 1992, raised from 15 to 145%; growth between 0.3 and 0.6 °C in the average global surface temperature, especially in mid-latitude continental areas; and global sea-level rise, between 10 and 25 cm, in the last 100 years.

As early as 2002, Ribeiro said that these arguments are sufficient to adopt at least three positions to address the issue to avoid the outbreak of conflicts between peoples and nations due to the consequences of climate change. First, they are the principle of common but differentiated responsibility, based on each country's history and considering economic development and fossil fuel use; the global ethics of becoming; and international environmental security.

From 2008 and 2009, in the IPCC reports filed in France and Denmark, the UN supported and ratified them as treaties to contain the effects of global warming. One is the United Nations Framework Convention on Climate Change (since 1994), which establishes a public commitment to reduce GHG emissions and recognizes the link between climate balance and biodiversity conservation. The other is the Kyoto Protocol, which advocates that developed countries are obliged to reduce their collective GHG emissions by at least 5%, compared to 1990 rates, for the 2008–2012 quadrennium—these levels have not been even remotely achieved! The decisive step for this protocol recognition was its ratification in Russia in October 2004. However, it only came into force on 02/16/2005, when the countries considered industrialized (36 of the 141 signatories) became responsible for reducing their combined GHG emissions.

Developing countries like Brazil, Mexico, and India had no targets for adoption—which generated skepticism from wealthy nations. The non-consideration of China as a developing country under the terms of the agreement, without obligation to the targets, motivated Canada's withdrawal in 2011. Rich and more polluting countries, such as the USA (non-signatories), defended compensatory policies, paying emerging countries by indexes of polluting gases that they do not release into the atmosphere, through Clean Development Mechanisms (CDMs)! The wealthiest countries are primarily responsible for these polluting gas emissions through technological use. Those under development are at the top of the *ranking* of emissions from biomass burning.

CDMs enable developed countries to meet their targets for reducing gaseous emissions by financing projects in developing countries. It is yet another command

and control instrument linked to the polluter-pays principle. They are "flexibility mechanisms" based on projects to reduce the emission of gases and capture carbon in the atmosphere to create a world carbon market, where 1 ton of gas corresponds to 1 carbon credit.

In 2016, the Paris Agreement, a more comprehensive substitute for the Kyoto Protocol, was adopted. It is considered a landmark among the most comprehensive and complete international climate agreements, with targets for developed and developing countries. Furthermore, the membership of nations was expanded with the participation of the countries above.

Developing countries with late industrialization, such as India and Brazil, which are among the six most polluting globally, are not committed to reducing emissions and were temporarily spared from this agreement. These countries prioritize social development, understood here as economic growth at any cost. In the case of Brazil, the current responsibility is the fight against deforestation to reduce the release of polluting gases, a silly measure.

The US is the world's biggest polluter, and its government argues that reducing pollution levels would strongly impact the national economy and stop its growth. However, the current government (Joseph Robinette "Joe" Biden) admits that combating climate change is indispensable. The most vulnerable populations to global warming would be the poorest on the planet, with a substantial impact on dryland regions and Areas Susceptible to Desertification (ASDs). However, this protocol established many proposals, including the creation of the Convention on Climate Change and its implementation conditions.

2.2 Climate, Global Warming, and Environmental Impacts

The conjunctural discussion on global warming is complex, not yet consensual, speculative, and, at times, polarized by economic groups and the media. Nevertheless, the IPCC and the majority of the scientific community appear as defenders of the thesis that there is enough evidence to speak of global warming caused by climate change arising from human actions—with which we agree.

Global warming, for several authors, can be linked to human or natural issues or even to both categories. In this regard, says Veríssimo (2003), IPCC (2018) and Gomes and Zanella (2021) when blaming human factors for such an event, they understand that GHG emissions into the atmosphere would be the factor that produces changes in the characteristics of this layer, so crucial for the earth's environmental balance, causing climate change.

On the other hand, about natural phenomena, it would be up to terrestrial movements and cyclical activities of the sun, directly affecting the solar energy *quantum* in the terrestrial atmosphere system, to trigger climate change by natural dynamics on deep time scales.

In this complex and controversial picture, Ayoade (2002) discussed some climatological aspects of the naturalness of global and regional climate changes. There are no simple unidirectional processes of cause and effect in climate but *feedback* in terms of an internal system (surface-atmosphere–ocean) or external (extraterrestrial factors). There are fluctuations and variations in the climate itself, which can follow trends or cycles and, over a long period, cause climate change.

A supposed climate trend may belong to a climate cycle. Local climate trends, cycles, or even changes can be asymmetrical to the global, regional, or continental pattern of climate fluctuations. It should not be forgotten that the diversity of terms used to describe climate variations (climate variability, climate fluctuations, climate trends, climate cycles, and climate change) is due to the appropriate and distinct time scales they validate. For example, climate variability can include fluctuations over periods of up to 35 years—therefore, very rapid—and cannot be considered climate change. There are also changes based on hundred years scales or instrumental (100–150 years), variations of thousands of years, or variations in the deep time scale, with durations in millions of years completes Ayoade (2002).

There are scientists and assessments, not signatories of the IPCC, arguing that no data or evidence confirms anthropogenetic climate change. It is said that this global warming lacks solid scientific bases and is mainly based on results from Climate Models (GCM), with mathematical equations that do not adequately represent the atmospheric physical processes, particularly the hydrological cycle. Future GCM projections, derived from hypothetical scenarios, would not be reliable and would not serve as a basis for planning human activities and social well-being. The natural variability is large and can be linked to the Pacific Decaden Oscillation (PDO), with a duration of 20 years in each episode. As a result, it is very likely that global cooling will occur in the next 20 years rather than warming, just to cite these discordant examples (Molion 2007, 2008a, b; FAO 2021; Baigorri and Caballero 2018).

However, the UN, through its multilateral bodies, such as the IPCC (and the World Meteorological Organization (WMO), through periodic reports based on assessments made around the world, stated in 2018 with a high level of confidence that human activities have caused 1.0 °C of global warming above pre-industrial levels, with a variation between 0.8 and 1.2 °C. Therefore, discussions now turn to the impacts related to this influence, its correlations, and ways to mitigate human societies that continue to inhabit the earth (Gomes and Zanella 2021; IPPC 2018, pp. 328–329).

Cyclical droughts, directly and indirectly, impact society, ecology, and the economy (Wilhite 2000; Albuquerque 2010; Shiferaw et al. 2014; Singh et al. 2014; Wilhite et al. 2014). Furthermore, the effects of natural drought conditions (and their impacts on society) will be further accentuated by the imminent threat of climate change. IPCC projections (2015, 2021) point to an intensification of this problem worldwide, both in intensity and frequency. In this role, the ASDs, with processes associated with desertification and affected rural communities, have their regional frameworks worsened (do Nascimento 2013; Bezerra et al. 2020).

The IPCC report (2018) projects the impacts of the increase in the global average temperature by 1.5 °C. Estimates that approximately 4% of the global will transform ecosystems from one type to another with global warming of 1 °C, compared with 13% at 2 °C for the next two decades. Some areas of the planet are expected to have landscapes and ecosystems not only impacted by climate change but even wholly

Table 2.2 Causes of climate change and indicators of past climates

Theories of causes	Indicators
Terrestrial causes: polar migration and continental drift; changes in the earth's topography; variations in atmospheric composition and distribution of continental and water surfaces, as well as snow and ice cover	Biological indicators: fossils, pollens and tree rings
Astronomical causes: changes in the eccentricity of the earth's orbit and in the precession of the equinoxes, as well as in the obliquity of the ecliptic plane	Lithogenetic indicators: annual layers of lake alluvium (varvites), evaporites, weathering processes, particularly laterization, and its products
Extraterrestrial causes: variations in the amount of solar radiation (solar *output*) and in the absorption of this radiation outside the earth's atmosphere	Morphological indicators: residual landforms (*inselbergs*, old beaches, dunes, and glacial landforms such as moraines and *eskrs*) and river terraces

Source Ayoade (2002) and Birth (2016)

transformed. The records of negative actions arising from human activity on nature set the tone for global changes with crucial impacts on the climate also originating in rural areas, highlighting: (i) intensive use of land in agriculture, especially monocultures; (ii) deforestation; (iii) incorrect land use and management; (iv) poorly planned and conducted irrigation; (v) rudimentary practices such as fires; (vi) wood extraction; (vii) disorderly urban sprawl, and so forth, (viii) water insufficiency related to irregular rainfall and excluding water distribution; (ix) indiscriminate use of wood to produce energy; and (x) extensive livestock and cattle trampling and non-adaptation of grazing vegetation (Ceará 2018, p. 237).

However, studies on climate change should gather the most outstanding sum of evidence from earth's geological time to the present time, and investigations of the past climate indicators, as shown in Table 2.2.

2.2.1 Global Changes and Climate Change: Resulting Desertification

Despite the environmental impacts of climate change and its consequences on desertification, simulation models of the general circulation of the atmosphere, in research by the *National Meteorological Center* (USA) for a period of ten years (1990–2000), showed that an increase in desert areas on the globe would cause a cooling of the troposphere in the tropics and subtropics, with a rapid weakening in the horizontal thermal gradient between the equator and the poles. As a result, there would be a reduction in total precipitation over deserts; this reduction would be 50% in Africa. In Southeast Asia and Australia, the decrease would be moderate—facts that have not yet been verified. With the growth of desert areas on the globe, precipitation would increase in Southwest North America, Equatorial Africa, the Tropical Atlantic, the

Indian Ocean, and the oceans to the West and Southwest of the desert areas of the Southern Hemisphere (Filho et al. 1994; Pereira and do Nascimento 2016).

In 1996, Williams and Robert Jr. highlighted the occurrence of these variations in different areas of the world on the interrelationship of climate change, climate variability, desertification, and economic losses and their international repercussions, in the domain of global climate change and its influence on desertification and desert formation. They mentioned the American West, the South of South America, South Africa, and the Australian dry regions, as well as the center of the Asian deserts, the Southwest of the USA, and, mainly, the Sahelian region, where significant changes are verified, with the reduction of rainfall since 1950.

With this scenario, climate variations and changes would be critical in regions with water deficits, such as in the drylands, influencing ecological and social adjustments, which will differ in temporal and spatial scale in the level and degree of changes.

The trend scenario of global changes has worsened due to changes in temperature and rainfall resulting from climate change. In this context, desertification is one of the significant and potential challenges with systemic risks that require more planning due to:

> (i) increase in soil temperature, which in turn will make some crops unfeasible due to the consequent reduction in soil moisture; (ii) increased water deficit; (iii) higher aridity index; (iv) soil salinization; (v) decline in agricultural production and productivity; (vi) loss of biodiversity (flora and fauna) generating systemic impacts such as pests and vectors; (vii) worsening social problems; (ix) return to the process of population migration to urban centers; (x) silting up of rivers and reservoirs; (xi) increase in edaphic droughts by decreasing soil water retention; and (xii) death of herds; among others. (Ceará 2018, p. 237)

As pointing deforestation out, their occurrence in the latitudinal ranges between 5° N and 5° S would cause an albedo increase and a mean global temperature reduction of 0.2–0.3 °C. In addition, changes in the surface-atmosphere system may reduce up to 10% of the evaporation and precipitation intensity at this latitude. Such events may be taking place in some regions of the world, hampering the life of societies that are not adopted or supported by governments—commonly negligent about these possibilities. For the drylands, the socioeconomic, cultural, and ecological damages can be even more harmful. At the environmental level, the following worrying factors are pointed out: the scarcity of water resources in the increasingly irregular drylands; the increase in the seasonal intermittency of the drainage network; lowering of the upper level of the water table; water sources with annual discharge, and post-season emission time increasingly shorter. These changes are worrisome, mainly associated with a decrease in rainfall in ASDs.

References

Albuquerque TMA (2010) Estudo dos processos de gestão de seca: aplicação no estado do Rio Grande do Sul, 425p. Tese (Doutorado em cursos Hídricos e Saneamento Ambiental)—Universidade Federal do Rio Grande do Sul, Porto Alegre

Ayoade JO (2002) Introdução à climatologia para os Trópicos, 8ª edn. Bertrand Brasil, Rio de Janeiro, 332p

Baigorri A, Caballero M (2018) Negacionismo, políticas demoscópicas y currículum de fracasos. El caso del cambio climático en España. Aposta Rev Cien Soc 77:8–58

Bensusan N (Org) Seria melhor mandar ladrilhar? Biodiversidade, como para quê, por quê. Ed. da UNB, instituto socioambiental, pp 13–28

Bezerra FGS et al (2020) Analysis of areas undergoing desertification, using EVI2 multi-temporal data based on MODIS imagery as indicator. Ecol Ind 117:106579

Ceará (2018) Governo do estado do ceará. Relatório: CEARÁ 2050, juntos pensando o futuro, Fortaleza

do Nascimento FR (2016) A desertificação como consequência da degradação ambiental. In: Barbosa e Ester Limonad JL (Org) Ordenamento territorial e Ambiental, vol 2, 2nd edn., Letra Capital, Rio de Janeiro, pp 267–292

FAO, IFAD, UNICEF, WFP and WHO (2021) The state of food security and nutrition in the World 2021. Transforming food systems for food security, improved nutrition and affordable healthy diets for all. Rome, FAO.

Filho JGCG et al (1994) Projeto Áridas: uma estratégia de desenvolvimento sustentável para o Nordeste. GTII. Recursos Hídricos: II.2 –Sustentabilidade do desenvolvimento do semiárido sob o ponto de vista dos recursos hídricos. Brasília, Ed do MMA, p 102

Furriela RB (2002) Mudanças climáticas globais e biodiversidade. In: Bensusan N (org.) Seria melhor mandar ladrilhar? Biodiversidade, como para quê, por quê. Brasília, Ed. da UNB, instituto socioambiental, pp 219–228

Gomes FIBP, Zanella ME (2021) Reflections on the natural and social impacts expected as a result of climate change in the Brazilian semi-arid. J Hyperspect Remote Sens 11:328–338

Intergovernmental Panel on Climate Change—IPCC (2015) Mudança do Clima 2014: Impactos, Adaptação e Vulnerabilidade. Available: http://www.iee.usp.br/sites/default/files/Relatorio_IPCC_portugues_2015.pdf. Acesso em: 13 Oct 2016

IPCC (2018) Special report. In: Global warming of 1.5 °C. Available: https://www.ipcc.ch/sr15/. Acesso: 5 Jan 2022

IPPC (2021) The intergonvernmental panel on climate change. Geneva. Cite on http://www.ipcc.ch, 02 Jan 2021

Molion LCB (2007) Desmistificando o aquecimento global. Intergeo, vol 5, pp 13–20

Molion LCB (2008a) Aquecimento global: uma visão crítica. Rev Brasil Climatol 3(4):7–24

Molion LCB (2008b) Mitos do aquecimento global, vol V. Plenarium Brasília, pp 48–65

ONU (2021) COP26. Cited on https://news.un.org/pt/tags/cop26. Acessado em 17 Jan 2021

Pereira JI, do Nascimento FR (2016) Susceptibility to desertification in Chicualacuala, Republic of Mozambique. Int J Geosci 07:229–237

Shiferaw B, Tesfaye K, Kassie M, Abate T, Prasanna BMM, Menkir A (2014) Managing vulnerability to drought and enhancing livelihood resilience in sub-Saharan Africa: technological, institutional and policy options. Weather Clim Extrem 3:67–79

Singh NP, Bantilan C, Byjesh K (2014) Vulnerability and policy relevance to drought in the semi-arid tropics of Asia—a retrospective analysis. Weather Clim Extrem 3:54–61

United Nations Convention to Combat Desertification (UNCCD) (2020) The great green wall implementation status and way ahead to 2030. Bonn, Germany, p 45

Veríssimo MEZ (2003) Algumas considerações sobre o aquecimento global e suas repercussões. Terra Livre, AGB - Rio de Janeiro, 20:137–144

Wilhite DA (2000) Drought as a natural hazard: concepts and definitions. In: Wilhite DA (ed) Drought: a global assessment. Routledge, London, pp 3–18

Wilhite DA, Sivakumar MVKK, Pulwarty R (2014) Managing drought risk in a changing climate: the role of national drought policy. Weather Clim Extrem 3:4–13

Chapter 3
Desertification: Concepts, Myths, and Reality

Abstract Difficulties remain to consider a consensual definition of desertification until today, even with the world climate crisis and increase in droughts. The difficulty in defining represents an obstacle to confronting the problem globally. Many ambiguities, criticisms, considerations, and disregards about desertification have been raised since the pioneering ideas on desertification in the middle of the last century to the formulation of its official definition at the United Nations Conference on Environment and Development (UNCED) or Eco'92. These aspects denote the complexity of the problem in its academic and scientific, political, social, cultural, environmental, fallacious, and sensationalist faces, in the face of global environmental problems, just to mention the most common.

Keywords Conceptual bases · Desertification · Desertization

3.1 Desertification × Desertization

The most common concept about desertification blamed the concomitant human and climatic factors for triggering the process. According to others, only the degradational socioeconomic factors of the land are relevant. In fact, desertification appeared to characterize the areas that were becoming similar to hot deserts or designate their expansion generically.

We realized, through the search to understand the official definition of desertification by the UN, that the concept was forged from contradictions and ambiguities, above all, to designate problematic areas in the Sahel—in areas that were the object of severe environmental degradation, allowing the expansion of the Saara's desert. Since then, this concept has been generalized across the globe and has spread the semantic misunderstanding between "desertification" and "desertization". The definition of desertification mentioned above occurs almost indistinctly by several scholars, politicians, civil society, and the technical and informational media, a fact that makes desertification quite permeated by jargon and clichés, demanding to be better understood.

F. Rodrigues do Nascimento, *Global Environmental Changes, Desertification and Sustainability*, SpringerBriefs in Latin American Studies, https://doi.org/10.1007/978-3-031-32947-0_3

Therefore, a distinction between "desertization" and "desertification" is mandatory because these terms are antinomic. We advocate the idea that in the drylands, the first term deals with the natural formation, expansion, or contraction of hot desert biomes, called physical-ecological, constituted over geological ages, independent of human action. On the other hand, desertification designates land degradation processes induced by socioeconomic activities, without necessarily having climate change, to the detriment of its widespread use that has become common sense (do Nascimento 2013, 2021).

These facts contribute to increasing the complexity and understanding of what desertification is. However, we need to delve a little further into this field.

3.2   The Complexity of Desertification Concepts and Their Understandings

In the investigated literature, 59 concepts about desertification were found. Some ambiguous, others generalist, without forgetting, of course, those that consider the official definition of the United Nations (UN); in all, there is a consensus that the aspects of degradation of natural resources, notably the drying of soils and the destruction of vegetation cover, vary from arid climatic ecozones to sub-humid dry ones.

Rubio (1995b) points out over one-hundred definitions of the desertification process, between the phenomenon's complexity, the biological concept of desert, and the multifunctional characteristics of the processes involved in desertification. In agreement with this point of view, Verdum et al. (2002) claimed the absence of dichotomy, but a hundred concepts related to the understanding and construction of the desertification concept and what would be done in the affected areas.

Initially, it is worth considering the evolution paths of the world debate about the phenomenon, expressing some conceptual contradictions, highlighting its current definition, and contributing to some policies to combat desertification developed until now. Then, some aspects of this phenomenon in Brazil and the world are discussed to understand better this problem in the Northeast and, specifically, in intermittent seasonal basins.

The problem of desertification is an old phenomenon, although it has only gained prominence in the last century. Historical reports show this problem in at least three regions in the world, which have incurred degradation/desertification processes for thousands of years: Mediterranean, Mesopotamia, and Chinese *Loess* (Dregne 1987). In Mesopotamia, the pioneering development of irrigation led to salinization, sodification/sodicity, and soil depletion in the lower reaches of the Tigris and Euphrates rivers, at least 2400 BC. Iraq, once a Mesopotamian territory, is experiencing environmental contingencies. It is a region depleted of renewable natural resources and, being arid and unproductive, a desert produced by socioeconomic activities (Conti 1994).

Another region affected is the eastern Mediterranean due to the excessive wood removal by the Phoenicians for use in civil and naval construction, freeing up land for agriculture. In addition, wood served as fuel in the iron smelting and supplied the Egyptian market to construct dwellings and temples around 2600 BC.

According to Rubio (1995a), there are references to the formation of barren areas in the code of Emperor Theodosius (438 AD), called *agri deserti* (literally: devasted areas); these would be areas abandoned by low annual productivity or by soil degradation promoted by military campaigns. In the nineteenth century, Chateaubriand (1768–1848), referring to the exhaustion of pedological potential in areas of the Iberian Peninsula, proclaimed the following sentence: "before civilization, the forest; later, the desert" (Rubio 1995a, p. 4).

Strictly speaking, the term desertification was coined in 1949 by Albert Abreuville, a French botanist and ecologist (Dregne 1987). The term characterized degraded areas of equatorial forests in former African colonies, due to the transfer of European scientific and technical knowledge and the misuse of natural resources, with rhexistasy developed from deforestation, causing the intensification of erosive processes and drying of soils, or desertification.

Although he was a pioneer in the study of desertification in the world, Abreuville did not give an exact and complete definition of the phenomenon, only conceptualizing it as the conversion of fertile lands into a desert from soil erosion linked to human activities. He studied humid and moderately humid tropical areas, observing processes of deforestation, irrational use of fire, and excessive cultivation that occurred in a period of occupation of 100–150 years.

According to Rubio (1995a), even before Abreuville, several authors, mainly English and French, reported severe ecological problems in Africa, Europe, Australia, and the United States, publishing significant titles that are now considered contributions to the study of desertification: *Man-made deserts*, from Lowdermilk, published in 1935; *Deserts on the March*, by Sears, edited in the same year; *The Rape of the Land*, by Jacks and Whyte, 1939; Vogt's *Road to Survival*, released in 1948; and the famous expression of HH Bennett, in 1939—*dust bowl*, that is: "clouds of dust".

However, the problem of desertification assumed worldwide proportions after World War II, notably with the environmental movement to protect nature. To enrich the discussion, the United Nations Educational, Scientific and Cultural Organization (UNESCO) launched its program for arid areas in Algeria in 1951. A year later, it sponsored new conventions on the subject, which led to the elaboration, in 20 volumes, of the series *Arid zones,* which gathers information about the arid ecozones of the globe.

In the wake of this initiative, the international geographic community, on the occasion of the XVIII International Congress of Geography, held in Rio de Janeiro in 1956, developed pioneering studies on the problem of desertification and arid lands. These studies had as their basic themes the use of the Negev desert, the mountain climate of the central Sahara, and the dry depressions of North Africa.

Undeniably, the significant droughts from 1934 to 1936, associated with environmental degradation, produced intense soil destruction in an area of 380 thousand km^2 in the states of Oklahoma, Kansas, New Mexico, and Colorado, in the

American Midwest, that caused the *dust bowl* phenomenon. Based on this episode, researchers studied the socioeconomic impacts on ecosystems marked by drought or water scarcity, becoming one of the main lines in discussing desertification and a comparative framework until today to study its occurrence.

Another serious environmental problem at the global level was the drying of the soils in the Sahel region, in Africa, after a severe drought that occurred between 1967 and 1973. This drought affected the vulnerable and already overexploited regional ecosystems, causing the death of 200,000 people and thousands of animals. Environmental degradation is a consequence of European colonization, enhancing environmental problems in this region. The fact that both phenomena were called desertification stands out.

The biggest problem in the Sahel was not the drought that occurred in the late 1960s (Mensching 1987). On the contrary, the drought only aggravated the degradation of biological resources resulting from the overexploitation of land and water, historically maintaining the headquarters of African colonies in Europe. Hence, an evident degradation of natural resources in the process of European colonization in Africa, highlighting serious problems of historical degradation that culminated in desertification processes. According to Rubio (1995b), desertification in the Sahel caused social and political conflicts, whose basis is the violent increase in demographic pressure on land use and unequal access to natural resources and regional wealth.

Regrettably, during the years marked by droughts, many Africans, especially children, died in conditions of hygienic and nutritional misery, while consumerist societies developed in the United States and Europe based on individual property and well-being, to the detriment of a socially egalitarian state of life.

In the last three decades of the twentieth century, however, discussions about the environmental degradation that transforms renewable natural resources into non-renewable ones, such as the drying of soils in arid, semi-arid, and dry sub-humid areas, generically took place under the prism of the problem of desertification.

Until then, the studies on desertification had been marked by USA and Sahel cases and always had ecological connotations, highlighted at the UN Environment Conference of Stockholm, 1972. On that occasion, 113 countries met, configuring a main historical event for the global environmental problems and representing the first moment of a global discussion on desertification. This event resulted in the report *Study on man's impact on climate*, which inspired the creation of the United Nations Environment Program (UNEP), promoting the discussion on desertification worldwide. Based on this conjuncture, environmentalism took shape and developed in the world.

Discussions on the topic began to be made by specific fields of knowledge. That same year, in Montreal, the International Geography Congress created a working group on desertification to compile a bibliography on the topic, encourage case studies, and develop thematic studies. At the next meeting of the International Geographical Union, held in Moscow in 1976, a symposium on desertification in arid lands was organized in Ashkhabad, then the Soviet Republic of Turkmenistan.

Although all this concern is relevant, desertification only began to be more carefully discussed at the UN Social and Economic Council in 1974, to hold a world congress to discuss the phenomenon in an interdisciplinary way in Nairobi, Kenya, three years later. The location was chosen strategically due to the deep soil drying caused by droughts between the 1950s and 1970s and the overexploitation of other natural resources.

In preparatory meetings for this event, coordinated by Mustafa Tolba, the problems of drying out soils were considered in the Sahara and several countries with areas of dry or sub-humid climate, removing the responsibilities of climatic contingencies as the leading cause for desertification. Thus, expressions such as "desert expansion" or "desert advance", which still resisted defining soil degradation in some regions of the earth, were elaborated from the term desertification.

At the Nairobi event, a produced text classified desertification into four axes from a global perspective, classified desertification, with specialists coordinating each one to provide greater conciseness and better organization of the debates: "Climate and desertification", under the responsibility of Here; "Ecological Changes", with Warren and Marzeis; "Population, Society and Desertification", by Kates, Johnson, and Haring; and "Technology and desertification", led by Garduño (Here et al. 1992).

With a progressive reinforcement of the need to debate desertification at the world level, international organizations, through the UN, began to create forums, congresses, and conventions to institutionalize the fight against this phenomenon. Brazil participated in several international events, including those of UNEP, to draw up a global action plan to combat desertification, according to Resolution No. in New York. This initiative preceded the holding, in August and September 1977, of the United Nations Conference to Combat Desertification (UNCOD) in Nairobi.

Since then, desertification has taken on a global and interdisciplinary character, recognized by UNEP as an environmental problem threatening the biosphere and triggering severe social costs (Rubio 1995b). The following concept was also accepted: arid or sub-humid ecosystems are impoverished due to the associative synergy of human activities and drought, and such changes can be measured by the decline in productivity and biological diversity, the increasing depletion of soils, and the risks driven by population contingents—that is, it is the decrease or destruction of the Earth's ecological potential that could definitively culminate in desert conditions.

The UNEP Secretariat encouraged Nairobi congress members to prepare illustrated documents on the accumulated knowledge, considering the latest information on the processes of desertification and its consequences for society and nature, as well as ways to combat them, denoting possibilities of reversing them. These documents addressed different issues, from physical-ecological to socio-technological, considering climatic, demographic, and socioeconomic aspects, desertification as an ecological change, and intervention technologies in arid and semi-arid zones. At the event, the motion: "Zero desertification by the year 2000" passed. However, twenty-two years after the deadline of that motion, the document did not have the desired effect even as global environmental changes advanced with the climate crisis.

As a result, the Action Plan to Combat Desertification (PACD) was created to develop actions worldwide, with voluntary adhesion of the countries that participated in the UNCOD. While the PACD recognized that the process could occur in any tropical, subtropical, and temperate water-stressed area, the geographic objective of the plan for the Nairobi Conference focused on the edges of all the world's hot deserts, encompassing areas where deserts were expanding, understood as desertification, and where it could occur, including semi-arid and sub-humid lands.

The PACD referenced the deserts of South, and North America, the Middle East, Iran, Israel, Pakistan, India, Central Asia, and China. A world map localized the deserts and areas at risk of desertification to verify the occurrence of such physiographic zones. This map represented the first moment of appreciation of the problem homogeneously and globally.

In general, a desert was considered to be any area of sparse or absent vegetation, and desertification is the expansion or intensification of such conditions. The UN General Assembly already used the expression "expansion and intensification of desert conditions". The PACD also focused on the human problems that influence desertification, even because inadequate socioeconomic activities fundamentally contribute to the genesis and aggravation of the process.

In the 1980s, the World Commission on Environment and Development (CMMAD) published the report *Our common future,* alerting to the fact that six million hectares of productive land suffered annually from the desertification process worldwide, and the widest soil loss was concentrated in Africa, the poorest continent on the planet.

Since the growth of desertification has been recognized as a threat to the human species, scientific expertise is the main component in reducing or reversing the associated impacts. As a result, knowledge about its primary causes was generated. Although the initial focus of discussion took place in Africa, it was in Latin America that this debate expanded, dominating the agendas of several congresses and conventions.

After UNCOD, in 1977, the International Conference on Climate Change and Sustainable Development in the Semiarid Region (ICID), held in Fortaleza, Brazil, in 1992, deserves to be highlighted. Desertification was considered holistically, associating it or not with climate change and, for the first time, an event focused on arid and semi-arid zones. In Brazil, this conference inspired the Guidelines for the National Desertification Control Policy. Project BRA 93/036. National Plan to Combat Desertification. ICID discussed the sustainable development of poor or developing countries, especially in the semi-arid regions of the tropical belt, including the Latin American Andes, Northeast Brazil, and the African Sahel, as well as the central regions of India and China.

This event was a world landmark, supported by 18 public and private institutions and the French, Dutch, and American governments, among others. Forty-five countries participated and gathered around 1000 participants—scientific community members, technicians, politicians, and representatives of various organizations, including non-governmental organizations linked to environmental issues.

As a result, ICID presented three essential documents: the Charter of Fortaleza, the Subsidies of the working groups, and the Declaration of the Secretaries of Agriculture of the states of the Brazilian Northeast. The latter expands the discussions on the problem of environmental degradation and climatic variations in semi-arid regions to support and legalize their demands with the strategy of, firstly, taking their results to New York, where, in March 1992, the last preparatory meeting for ECO-92.

Furthermore, ICID is for the world what CONSLAD is for Latin America and the Caribbean. Indeed, that conference was decisive, bringing new global paradigms.

Having achieved this objective, the formulations built in the USA became proposals for discussions about degradation/desertification, and this theme was included as an agenda for debate. In addition, ICID put this issue on the agenda at ECO-92, generating investments to address this problem worldwide.

More than a scientific meeting for the exchange of ideas, ECO-92 was a conference of States, politically and diplomatically permeated by their interests. Consequently, industrialized countries strongly influenced its agenda, with discussions focused on global issues specific to humid regions, such as the Amazon, relegating the problems of soil drying outside these physiographic zones to a secondary level.

Prioritizing topics such as global environmental changes, the climate crisis, the depletion of the ozone layer, and biodiversity protection, that convention on environment and development underestimated central issues for poor or developing countries, such as the deep degradation of natural resources leading to desertification.

However, several reports from nations to prepare for the United Nations Conference on Environment and Development (Cima) dealt with soil degradation, widespread deforestation, fires, siltation of water bodies, and mining, among other environmental problems, not expressly as desertification. However, in the United Nations Convention to Combat Desertification (UNCCD or, in short, CCD), held in Paris, 1994, desertification was ratified by 196 countries, mainly in Africa, serving as a framework for the institution of the World Day to Combat Desertification.

Continuing with ICID and UNCTAD discussions on desertification, the National Conference and Latin American Seminar on Desertification (CONSLAD) was held in Brazil (Fortaleza city). Its objective was to contribute to better political positioning of Latin American countries in a joint effort to occupy spaces in the negotiations on desertification, in terms of obtaining resources, in addition to annexing the North of the Mediterranean and Asia, not only privileging the African continent.

3.3 The Institutionalization of the Desertification Phenomenon

The desertification process was included in ECO-92 and the wake of global environmental changes, considering the events that have been related so far. Agenda 21 finally provided an official definition of this phenomenon in its Chap. 12.2, giving priority to the desertification prevention for degraded or slightly degraded lands without neglecting seriously degraded ones and encouraging the participation of local communities, governments (national and regional), and NGOs with the aim of sustainable development (Pereira and do Nascimento 2020; UN 2022; UNESCO 2022). Thus, in the 1990s, the Nairobi Conference's indication that desertification is a problem of the first magnitude was ratified, and the Convention incorporated the precepts of this agenda to Combat Desertification.

This agenda, the UN Convention to Combat Desertification, 1994—an offshoot of the Nairobi discussions, 1977, relates desertification to climate data. The prevention, correction, and regeneration of desertified areas is a function of investment in (re)forestation. The population's participation is also considered essential, especially for appropriating local knowledge and creating gene banks to increase biodiversity. In this plot, different currents of opinion question the concrete occurrence of the desertification process. Formally, however, through that document and the Convention to Combat Desertification (CCD), the UN component countries consolidated the idea that desertification is a problem of degradation in drylands.

The highlight should also be given to the II International Conference on Climate, Sustainability and Development in Arid and Semi-arid Regions (ICID + 18), held in Fortaleza, Ceará, in August 2010. It was the most recent and the largest of all events carried out by COMLAND since 1972, the beginning of world discussions on desertification.

An important issue is the differentiation made between desertification and drought. UNCOD defines drought as a natural phenomenon triggered when precipitation is significantly reduced from its average level, causing severe hydrological imbalances and negatively impacting dryland production systems. In this context, there is also the controversial issue of the greenhouse effect and its consequences, especially concerning climate change. In this way, the long-term effects of desertification are different from the effects of droughts, which are to some extent transient.

The CCD considers dry semi-arid and sub-humid areas as all those with aridity index between 0.05 and 0.65, except for polar and sub-polar areas. The same entity described affected zones as those arid or dry sub-humid areas affected or threatened by desertification (Brasil 2004). Regarding the causes and complexities of approaching desertification, we highlight some works that cross information in this regard. More information can be found in Chap. 2 and the bibliographic references below.

References

ABC de las Naciones Unidas (2012) United Nations Department of Public Information, Paris, 449p. Cited on https://unesdoc.unesco.org/. Acessado: 16 Jan 2022

Brasil/Ministério do Meio Ambiente (MMA) (2004) Programa de Ação Nacional de Combate à Desertificação e Mitigação dos Efeitos da Seca, PAN-BRASIL. Edição Comemorativa dos 10 anos da Convenção das Nações Unidades de Combate à Desertificação e Mitigação dos Efeitos da Seca—CCD. MMA, Brasília, 225p

Conti JB (1994) O conceito de desertificação. In: Anais do 5 Congresso Brasileiro de Geógrafos: Velho mundo – novas fronteiras: perspectivas da Geografia Brasileira. AGB, Curitiba – PR, 1:366–368 do Nascimento FR (2013) O Fenômeno da Desertifcação, vol 1, 1st edn. EDITORA UFG, Goiânia, 240p

do Nascimento FR (2021) Desertificación—Diccionario de desarrollo regional y cuestiones conexas-. Diccionario de Desarrollo Regional y Cuestiones Conexas, vol 2, 2nd edn. Editora Conceito, Uruguaiana, pp 245–249

Dregne HE (1987) Envergadura y difusión del processo de desertificación. In: Programa de las Naciones Unidas para el Médio Ambiente (PNUMA): Comision de la URSS de los Asuntos de PNUMA. Colonizacion de los territórios áridos y lucha contra la desertification: enfoque integral. Moscu: Centro de los Proyectos Internacionales—GKNT, pp 10–17 (Capitulo I)

Here FK et al (1992) Desertificação: uma visão global. In: Desertificação causas e conseqüências. (Trad. Port) Lisboa, Fundação Caloute Gulbenkina, pp 12–108

Mensching HG (1987) Desertificacion em la zona de Sahel. In: Programa de las Naciones Unidas para el Médio Ambiente (PNUMA): Comision de la URSS de los Asuntos de PNUMA. Colonizacion de los territórios áridos y lucha contra la desertification: enfoque integral. Centro de los Proyectos Internacionales—GKNT, Moscu, pp 72–76 (Capitulo XI)

ONU (2022) Agenda 21. Cited on Agenda 21. https://news.un.org/pt/tags/agenda-21. Acessado em 18 Jan 2022

Pereira JI, Nascimento FR (2020) Focusing on the susceptibility to desertification in chicualacuala, Republic of Mozambique. In: Huan Yu (Org) Focusing on the susceptibility to desertification in chicualacuala, Republic of Mozambique, 1st edn. Book Publisher Internacionational, Inglaterra, 5:18–29

Rubio JL (1995a) Desertification: evaluation of a concept. In: Seminário Desertificación y Cambio Climático. Centro de Investigaciones sobre Desertificación—CIDE/Universidad internacional Menendez Pelayo (UIMP), C.S.I.C—Valencia, 9p

Rubio J (1995b) Definiciones. Marco conceptual. In: Seminário Desertificación y Cambio Climático. Centro de Investigaciones sobre Desertificación—CIDE/Universidad Internacional Menendez Pelayo (UIMP), C.S.I.C—Valencia, 46p

United Nations (2022) ABC de las Naciones Unidas. United Nations Department of Public Information. Paris 2012, Cited on https://unesdoc.unesco.org/. Acessado: 16 Jan 2022, p 449

UNESCO (2022) Um glossário para o antropoceno. Cited on 17 Feb 2022

Verdum R et al (2002) Desertificação: questionando as bases conceituais, escalas de análise e conseqüências. In: GEOgraphia, Revista da Pós-Graduação em Geografia do Departamento de Geografia da Universidade Federal Fluminense. Ano 3, n° 6, UFF/EGG, Niterói, pp 119–132

Chapter 4
Causes and Impacts of Desertification in the World

Abstract The causes and repercussions of desertification worldwide vary in type, origin, degree, and scope. As for the origin, desertification can depend on physical or human factors. Its magnitude corresponds to the degree of intensity of the phenomenon and its related effects. Comprehensiveness has to do with its scale and scope. The leading causes of desertification are (1) traditional agriculture, undercapitalized and with a low technological level; (2) irrigated agriculture, capital intensive but poorly managed; and (3) contingencies of climate change and extreme events. Among the leading causes of desertification in non-irrigated areas in the world are population growth and lack of associated policies; income concentration and social exclusion; increase in livestock; overgrazing of cultivated native pastures and agricultural use by crops; fundamental sanitation problems and irrational farming techniques; and plant and mineral extraction. Soil salinization is the primary agent causing this problem in areas with irrigated agriculture. Moreover, the consequences of climate change represent the contingencies of intensive use of natural resources and the non-neutrality of CO_2 emissions—elements that cause environmental degradation and desertification. Structural factors, however, such as income concentration and inadequacy of some economic, cultural, and technological activities to environmental conditions, make it difficult to contend with desertification and magnify the effects of the causes above, commonly in a social context where poverty is marked. In this way, more than climatic vicissitudes, human interference in the biophysical environment causes a rupture in the dynamics of landscapes, mainly due to the degradation of vegetation and biological and climato-hydrological elements.

Keywords Causes · Consequences · Environmental degradation · Climate change

4.1 Effects and Consequences of Desertification

Although many effects of desertification are systemically correlated with global, regional, and local climate factors, Drew (1986) lists some of these effects in the following Box. One of the great difficulties in desertification studies is that this

© The Author(s), under exclusive license to Springer Nature Switzerland AG 2023
F. Rodrigues do Nascimento, *Global Environmental Changes, Desertification and Sustainability*, SpringerBriefs in Latin American Studies,
https://doi.org/10.1007/978-3-031-32947-0_4

phenomenon results from the simultaneous action of several causes in a determined area on different space-temporal scales. The scale issue is essential to differentiate the processes acting at the local level, such as erosion, salinization, and crusting of the soil, as well as the decrease in its organic matter.

There is enormous complexity in understanding the causes of desertification, which stem from climatic or human effects, from the climatic rigors related to rainfall to the standard of living and the pressure of human populations, through the level of development of nations and their prevention policies to degradation and droughts (Table 4.1).

Dregne (1987) believes that the factors of propensity to desertification in Africa have the same profile verified in Asia and Latin America and can be listed in three groups: population growth and increase in the number of cattle, problems of basic sanitation, and irrational agriculture techniques. In particular, plant and mineral extraction, overgrazing of native or cultivated pastures, and agricultural crop use can be the leading causes in non-irrigated areas. On the other hand, soil salinization is the primary agent in areas with irrigated agriculture (CNRBC 2004).

Zoon and Orlovski (1987) and Castro et al. (2019) say that alkalization and secondary salinization, caused by irrigation, are recorded in history as anthropogenic factors of desertification, as well as chemical contamination from agriculture, with the

Table 4.1 Effects of desertification

Factors affected	Direction of change	Immediate effects	After-effects and feedback
Slightly affected			
Solar radiation	Any less dust	Surface cooling	Atmospheric cooling, less convection, less rain, less vegetation, albedo—greater cooling (positive feedback)
Moderately affected			
Albedo Subsurface flow Percolation of groundwater	Bigger Smaller Smaller	Surface cooling; less water for the sources Lowering of the water table; wells dry up	Positive feedback with depopulation—lower voltage (negative feedback); decrease population, livestock, and grazing; more vegetation, soil improvement, more infiltration (negative feedback)
Severely affected			
Vegetation Evaporation Infiltration Surface runoff	Smaller Bigger Smaller Bigger	Erosion, depopulation Decline of vegetation Idem Erosion	Negative feedback and/or cooling—drought intensification (positive feedback) Idem Idem Less groundwater, changing landforms, less vegetation, etc.

Source Drew (1986)

intensification of production through fertilizers, insecticides, herbicides, and other chemicals, which can contaminate animals and humans alike.

According to Pernambuco (2001), ultimately, the causes of desertification are associated with two types of problems related to traditional agriculture, which is undercapitalized and with a low technological level, and irrigated agriculture, which is intensive in the capital but poorly managed. In both cases, the cause of this environmental deterioration is the growing market pressure that affects these areas. However, structural factors, such as income concentration, high population density, and inadequacy of some economic activities to environmental conditions, hamper the opposition to desertification and amplify the effects of the causes above. Considering the multiplicity of factors that generate desertification, particular emphasis was given to the leading global causes of this problem, as presented in Table 4.2.

In qualitative terms, the *International Center for Arid and Semi-Arid Land Studies—ICASALS*, from the University of Texas, states that degradation affects 69% of all land worldwide since it includes, analytically, areas that present any vegetation degradation, even if the soils are not degraded.

Based on the official definition of desertification, UNEP considered the susceptible areas based on zonal classification and climate classes defined by the aridity index. Strictly speaking, the definition of aridity based on the ratio between precipitation and potential evapotranspiration (P/ETP) was established by the UN in 1977 in its Action Plan to Combat Desertification, published in the work *Map of the world distribution of arid regions*, prepared by UNESCO in 1979, which considers areas susceptible to desertification to be those with arid, semi-arid, and dry sub-humid climates.

Currently, the aridity index is better known as the Thornthwaite formula, from which UNEP created the *World Atlas of Desertification* (Fig. 4.1), defining risk areas and serving as a global parameter with the establishment of the following climate classes: hyperarid < 0.03; arid 0.03–0.2; semi-arid 0.21–0.50; dry sub-humid 0.51–0.65; and humid sub-humid > 0.65 (i.e., there is no aridity). Concerning areas susceptible to desertification, the framework indicates very high susceptibility 0.05–0.20; high 0.21–0.50; and moderate 0.51–0.65. This indicator means that, *roughly speaking*, the drier the area, the more susceptible it will be to desertification (Brasil 2004a).

Deepening the issue, arid, semi-arid, and dry sub-humid zones (Drylands) correspond to those in which the proportion between average precipitation/year and Potential Evapotranspiration (TEP) is greater than or equal to 0.05 or less than 0.65, and the annual mean temperature exceeds 0 °C.

The degree of aridity depends on the amount of water coming from rainfall (P) and the maximum water loss through evaporation and transpiration, or evapotranspiration (ETP). It is important to note that the aridity index was changed in its last class, from sub-humid (0.50–0.75) to dry sub-humid (0.51–0.65), because the sub-humid areas hold greater biodiversity than their dry derivation. The concept of desertification as a decrease in the biological potential of the land was then ratified and specified by the World Meteorological Organization, using the aridity index.

Based on this index, the official definition of desertification excludes the hyperarid regions of the world, such as the Atacama and Sahara deserts. However, it advances

Table 4.2 Main causes of desertification in the world

Zonn and Orlovski	Rubio	Rubio and Bochet	Redesert	Pan	Pereira and do Nascimento	Bezerra et al.	UN	
								Changes in the patterns of socio-spatial organization resulting from the intensification of production in response to market integration processes at the regional, national, or international level with the current conjuncture of globalization. Market integration, technological development indices, land distribution, disorderly urban expansion, destruction of vegetation, intensity and management of natural resources use, inappropriate agricultural and livestock practices, and socioeconomic effects of climate variability
			X	X	X	X	X	Cultural characteristics and degree of economic development of populations. Poor areas are more vulnerable. In large part, desertification occurs due to poverty, causing food insecurity associated with variations in the hydrological cycle, such as droughts and floods

(continued)

Table 4.2 (continued)

Zonn and Orlovski	Rubio	Rubio and Bochet	Redesert	Pan	Pereira and do Nascimento	Bezerra et al.	UN
							Changes in the patterns of socio-spatial organization resulting from the intensification of production in response to market integration processes at the regional, national, or international level with the current conjuncture of globalization. Market integration, technological development indices, land distribution, disorderly urban expansion, destruction of vegetation, intensity and management of natural resources use, inappropriate agricultural and livestock practices, and socioeconomic effects of climate variability
			X	X	X		Lack of productive articulation between agriculture and livestock, intensive cultivation and grazing, deforestation, mining, and inadequate irrigation practices
	X	X			X		Destruction of biological exploitation factors by road construction, industries, mining, geological exploration, and irrigation works

(continued)

Table 4.2 (continued)

Zonn and Orlovski	Rubio	Rubio and Bochet	Redesert	Pan	Pereira and do Nascimento	Bezerra et al.	UN	
								Changes in the patterns of socio-spatial organization resulting from the intensification of production in response to market integration processes at the regional, national, or international level with the current conjuncture of globalization. Market integration, technological development indices, land distribution, disorderly urban expansion, destruction of vegetation, intensity and management of natural resources use, inappropriate agricultural and livestock practices, and socioeconomic effects of climate variability
X	X	X		X	X	X	X	Growth of saline deserts in endorheic basins
X	X					X	X	Biological degradation, especially of vegetation cover, with reduction of organic matter. Physical degradation, with adverse changes in soil properties—porosity, bulk density, structural stability, and permeability. Factors such as salinization, acidification, contamination, leaching, alkalization, and flooding of irrigable or surrounding lands are added

(continued)

Table 4.2 (continued)

Zonn and Orlovski	Rubio	Rubio and Bochet	Redesert	Pan	Pereira and do Nascimento	Bezerra et al.	UN	
		X		X	X			Changes in the patterns of socio-spatial organization resulting from the intensification of production in response to market integration processes at the regional, national, or international level with the current conjuncture of globalization. Market integration, technological development indices, land distribution, disorderly urban expansion, destruction of vegetation, intensity and management of natural resources use, inappropriate agricultural and livestock practices, and socioeconomic effects of climate variability
					X	X		Inadequate management of natural resources and climate change affect food production on land

Source do Nascimento (2006, 2021), Zonn and Orlovski (1987), Rubio (1995a), Rubio and Bochet (1998), Redesert (2003), Brasil (2004a, b), Pereira and do Nascimento (2016)

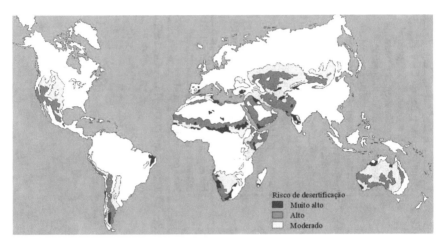

Fig. 4.1 Areas at risk of desertification. *Source* United Nations Conference on Desertification (1977)

by breaking with a purely climatic vision, considering environmental degradation not only soil erosion but also a social and economic problem, and underlining the delimitation of ecozones susceptible to the application of the CCD. On the other hand, this climatological framework is restricted to the dry semi-arid and sub-humid regions of the world, favoring African countries. Moreover, the framework does not consider other areas which show signs of deep environmental degradation: for example, Alegrete, in southern Brazil, which has an aridity index above 1.0 but has been suffering severe environmental degradation with the drying of the soil.

The most important thing was that UNEP recognized, even before the Nairobi meeting, that degradation/desertification can occur in any tropical, subtropical, and temperate region—in semi-arid and dry sub-humid areas climates—regardless of regional climate changes. The UNCOD, in 1977, pointed out issues involving socioeconomic and demographic aspects, in addition to ecological changes. Therefore, the aridity index must always be contextualized with factors causing desertification to avoid reductionism and climate determinism.

Mainguet (1995a, b), quoted by Verdum et al. (2002a, b) and Verdum (2004) established five phases to define better the conceptual variations regarding desertification in its different spatiotemporal scales:

(1) Awareness phase—Human activities as the main causes of desertification—soil and vegetation deterioration. Abreuville, a precursor of this awareness, pointed out, in 1949, the emergence of "true" deserts in the former European colonies in North Africa, where rainfall varies from 700 to 1500 mm/year. The European technical-agricultural changes applied in Africa triggered the desertification processes.

(2) Phase of the exaggerated perception of the process—From 1970 onwards, with the need to develop a single internationally accepted concept by using remote

sensing and climate data for mapping on a planetary scale. UNCOD scientifically recognizes the process.

(3) Doubt about the process—End of the 1980s. Characterization of desertification: its extension, causes and solutions, irreversibility or not, and complex dynamics of the sands about irradiating centers with population concentration. Doubts about the profitability of the applied technologies are added.

(4) The myth of desert growth—Generalization of desert growth. Desertification was characterized mainly by punctual environmental degradation around the villages than by the linear increase of the margins of the great deserts.

(5) New realism on desertification—He considers knowledge of climate dynamics insufficient to indicate the causes of rainfall and, consequently, hydrological scarcity, especially in the Sahel, between 1968 and 1985. This idea, in a way, persists to this day.

However, in certain ecozones, such as the dry sub-humid areas, there is an erroneous view about the degraded or regenerated geographic space, with or without human intervention. Furthermore, skepticism about the growth of deserts persists in common sense, based on remote sensing work to ascertain the extent of this degradation and the regenerative capacity of the natural environment. For a general contextualization of the historical events that contributed to the CCD, Table 4.3 shows the antecedents of events related to the fight against desertification.

Desertification, in this way, comes to be understood as a singular process, combining local variables, and no longer as a generalized process of degraded ecozones. Therefore, it is imperative to study its causes, processes, and effects in the search for the determination of the combinations of active variables, avoiding generalizations, jargon, clichés, and indistinct uses of the term desertification. The media are potential vehicles for this. However, although they publicize the problem, they do so in a way that the population has a simplistic reading of the phenomenon since the term "desertification" applies to any area, of any geographical scope, in a declining process of the productive biological environment (Rubio 1995b). Therefore, a distinction will be made between desertification and desertization as an environmental phenomenon (physical and human) and strictly physical-ecological.

Conti (2003) says that desertification has two modalities: natural (or climatic) and human (or ecological). The author also defines desertization as the "extension of typically desert landscapes and forms, in semi-arid or sub-humid areas, as a consequence of human action". For him, "desertification" is a term of imprecise meaning and without consensus among scholars, who use it less and less. Other terms have already been proposed, such as "Sahelization", "Stepization", and "Sudanization", corresponding to different stages of disappearance of tree capping; and also "aridization" or "aridification", to indicate a natural evolution toward a drier climate. The reverse occurrence, when there is the retreat of deserts, is called anti-desertification.

In this context, Nimer (1988) says that a desert is a function of extreme climatic aridity, independently of human action. When desertification is a process of evolution toward a given desert or based on it, this author considers this expression of inappropriate use for cases that do not have a cause-and-effect relationship with regional

Table 4.3 Historical background of events related to combating desertification

Fact	Date/Period	Local
Severe and extensive drought ravages the US	1930s	US
Drought in Africa, exacerbated by environmental devastation that impacted agricultural production and productivity, leading to the impoverishment of populations. Rising infant mortality rates, epidemics, famine, and war caused international commotion, due to the migration of a legion of refugees	1960s, especially from the 1970s	Sub-Saharan Africa
First International Conference on the Human Environment, promoted by the UN. Discussion of numerous topics relevant to the environment, including the African catastrophe, represented by the great drought of the Sahel (1967–1973)	1972	Stockholm, Sweden
Creation of the Interministerial Standing Committee to Fight Drought in the Sahel (CLISS). It was attended by representatives from the Sahel region	September 1973	Africa
Brazil begins preparations for the national report to be presented, in 1977, at the conference on desertification convened by the UN	1974	Recife, Brazil
Publication of the book *The great Brazilian desert*, by Vasconcelos Sobrinho, and creation of the concept of a desertification nucleus, officially adopted until today	1974	Recife, Brazil
First UN conference on desertification, in which this problem was recognized worldwide. Brazil presented to the world its situation regarding this problem, until then only recognized as serious in Africa	August–September 1977	Nairobi, Kenya
Embrapa (CPATSA), Fundação Joaquim Nabuco, UFRPE, Sudene, and Núcleo Desert at UFPI develop studies and works related to the Brazilian semi-arid region	1980s	Brazil
UNEP evaluated the actions taken, noting the poor performance of the actions of the first conference. With these results, several countries with desertification problems, especially in Africa, decided to propose a convention on the subject	1991	Nairobi, Kenya
Holding of the only world event dedicated to the arid and semi-arid regions of the planet, ICID, preparatory to UNCTAD (ECO-92), bringing together representatives from more than 70 countries on four continents. Focusing on the countries affected by desertification, they consolidated their technical and political bases to demand the celebration of a specific convention for these areas	January–February 1992	Fortaleza Brazil

(continued)

Table 4.3 (continued)

Fact	Date/Period	Local
UNCTAD (ECO-92). Brazil stands out in the discussions on desertification, which resulted in the negotiation of a convention to combat desertification, proposed by African countries in compliance with what is recommended in Chapter 12 of the global Agenda 21, launched and approved at this conference. NGOs from five continents, gathered in the forum parallel to ECO-92, prepared 46 agreements, including the Treaty on Arid and Semi-arid Zones	June 1992	Rio de Janeiro, Brazil
Creation of the intergovernmental panel to negotiate the text of the convention	June 1992	New York, USA
Meeting to discuss the preliminary text of the convention. Brazil and Latin America had the participation of government and civil society representatives	August 1993	Brasilia Brazil
CONSLAD, in which government and civil society representatives from Latin America formulate and negotiate the final text of *the Latin American Regional Annex*	February 1994	Fortaleza, Brazil
Latin America Regional Annex is approved by the intergovernmental panel of negotiations. The original text of this document serves as a basis for the negotiations on the regional annexes for Asia and the North of the Mediterranean	March 1994	Geneva, Switzerland
Conclusion of the UNCCD negotiations, or simply CCD. The date of June 17 was consecrated as the World Day to Combat Desertification	June 17, 1994	Paris, France
Adhesion of Brazil to the CCD, in a formal act of its government	October 15, 1994	Brasilia, Brazil
Agreement between the federal government and the UNDP and FAO, aimed at drawing up the National Program to Combat Desertification—PAN-Brazil	1994–1998	Brasilia, Brazil
The convention comes into force after ratification by 50 countries	December 26, 1996	New York, USA
The Brazilian National Congress ratifies the CCD	June 25, 1997	Brasilia, Brazil
The CCD becomes effective in Brazil	September 24, 1997	Brasilia, Brazil

(continued)

Table 4.3 (continued)

Fact	Date/Period	Local
Conference of the Parties to the United Nations Convention to Combat Desertification (COP 1), with attention to bureaucratic and financial issues related to the functioning of the CCD. Rules for the functioning of the COPs were established, and the functions of the global mechanism responsible for financing the convention were regulated	September 1997	Rome, Italy
Kyoto Protocol—more rigid commitments to reduce the emission of gases that produce the greenhouse effect, which is the cause of the current global warming	12/11/1997	Kyoto, Japan
COP 2. Different technical aspects were prioritized, such as desertification indicators, traditional knowledge, and information network. Holding of the first parliament meeting on desertification	November 1998	Dakar, Senegal
COP 3. Preparation of medium-term goals, to be met by the CCD, and definition of forms of operation and activities of a global mechanism	November 1999	Recife, Brazil
Preparation of *the semi-arid declaration*, during COP 3, through a parallel forum promoted by civil society. Consolidation of Articulation in the Brazilian Semi-Arid Region (ASA), the largest organization of the national society for coexistence with the semi-arid region	November 1999	Recife, Brazil
COP 4. Adoption of an annex on the accession of Central and Eastern European countries, providing for the 2001–2010 decade the potential for carrying out activities to combat desertification and mitigate the effects of drought	December 2000	Bonn, Germany
COP 5. Special mention to the Science and Technology Committee, which defined the parameters and indexes for the alert systems to combat desertification. Establishment of a group of experts to support the technical examination of issues related to desertification and creation of the Review Committee on the Application of the Convention (CRIC)	October 2001	Geneva, Switzerland
Creation of a Working Group to Combat Desertification—ASA, GTCD/ASA—to articulate civil society actions on the issue of desertification	April 2002	Recife, Brazil
World Summit on Sustainable Development. On this occasion, governments appealed to the Global Environment Facility (GEF) to become a CCD financing mechanism	August/September 2002	Recife, Brazil

(continued)

Table 4.3 (continued)

Fact	Date/Period	Local
Agreement between Fundação Grupo Esquel/Brasil, IICA, the Inter-American Development Bank (IDB), and the special fund of the government of Japan to implement the Program to Combat Desertification and Mitigation of Drought in South America, involving Argentina, Bolivia, Brazil, Chile, Ecuador, and Peru	September 2002	Brasilia, Brazil
First meeting of the Convention Implementation Review Committee (CRIC). Agreement signed by the countries participating in the IDB-IICA-FGEB agreement	November 2002	Rome, Italy
National Meeting on Desertification of the Brazilian Semiarid Region, in which, for the first time, representatives of civil society entities, from desertification centers, met to discuss the topic	June 2003	Pernambuco, Brazil
IX CCD Regional Meeting, in which Brazil is elected as the Southern Cone representative for the regional executive committee, which aims to collaborate in the coordination of CCD application activities in Latin America and the Caribbean. At this same meeting, the Regional Network on Desertification and Drought in Latin America and the Caribbean Region—Deselac—was re-implemented	June 2003	Bogota, Colombia
COP 6. Definition of the GEF as the convention's financial mechanism. Increase in the budget of the global mechanism, the convention's financing instrument	August/September 2003	Havana, Cuba
PAN-Brazil participatory elaboration process	June 2003 June 2004	Brazil
Launch of PAN-Brazil at the South American Conference on Combating Desertification. Largest institutional framework on the subject in the country	3 August 6, 2003	Fortaleza, Brazil
COP 15/CMP 5 and Intergovernmental Panel on Climate Change	November and December 2009	Copenhagen, Denmark
Decade on Deserts and Combating Desertification. Official announcement made at the opening of the II International Conference on Climate, Sustainability and Development in Arid and Semi-arid Regions	August 16, 2010	Fortaleza Brazil
United Nations Conference on Sustainable Development (UNCDS)—Rio+20, which discussed the renewal of political commitment to sustainable development	13 to 2/06/2012	Rio de Janeiro, Brazil
Paris Agreement: GEF emission reduction measures from 2020 onwards in order to contain global warming below 2 °C, preferably 1.5 °C, and strengthen countries' capacity to respond to sustainable development. The agreement was negotiated during COP21	12/12/2015	Paris

(continued)

Table 4.3 (continued)

Fact	Date/Period	Local
26th Conference of the Parties to the UN Framework Convention on Climate Change 2021 (COP26). Added the 15th meeting of the parties to the Kyoto Protocol (CMP16) and the 2nd meeting of the parties to the Paris Agreement (CMA3)	1 and 11/12/2021	Glasgow, Scotland

Source Brasil (2004b) and do Nascimento (2013) and Pereira and do Nascimento (2020)

climate change—which still could not be proven—similarly to what happens in the Sahel surroundings. For these cases, Nimer suggests the term semi-desertification.

Contrary to Conti, Rubio (1995b) reserves the word "desertization" exclusively to designate the natural processes of desert formation, the so-called physical-ecological deserts; that is, those "processes of expansion and contraction of desert areas on geological time scales". In turn, the term "desertification" was conceived to designate human-induced land degradation processes. This notation was adopted in this work, which is more precisely called environmental degradation/desertification. However, Le Houvérou (1989), cited by Suertegaray (2003), says that the use of the term "desertization" in Europe is inappropriate, as there are no deserts on this continent, except for Southeast Spain and the Middle Ebro Valley. Furthermore, on the other side, in the last 300 years, there has been an increase in reforestation across the continent.

Verdum et al. (2002a, b) claim the existence of two distinct time scales: the geological and the human, reflecting the variability of factors and scales in the generation of desertification in its understanding by the scientific world, which (re)produces this variability to the detriment of consensus about the concept, according to. Therefore, even if desertification is not conceived exclusively as a global process, proclaims this author, it must be considered in its local specificities.

However, in a reductionist way, many treat desertification only as soil degradation when, in reality, what can happen is the indistinct use of this concept when dealing with environmental problems related to the depletion of the productive capacity of renewable natural resources.

According to Verdum et al. (2002a, b), in the example of Uruguay, soil degradation considered as desertification serves as an argument for capitalization of resources, which demonstrates the opening of a wide field of possibilities. In another South American country, Brazil, physical-ecological desertification does not occur. However, the implications of uninhabitability and emerging impacts may be similar, especially concerning the drying of soils and the withering of water resources.

After almost three decades, UNEP, evaluating the PACD, concluded that the results obtained were modest since the national actions of countries with serious environmental problems, as well as human training for the optimization of human resources, was not developed by the world.

On diversity and distortions in the understanding of the desertification process, Verdum et al. (2002a, b) believe that, from the proposal to combat it, there are controversies in the discussion of the scale (temporal or spatial) in which they are considered, in addition to climatic, botanical, pedological, and social aspects, to mention the most relevant. In turn, Suertegaray (2003) argues that the concept is controversial due to the complexity of the causes of the phenomenon. Table 4.4 displays the theme complexity, trying to outline some differences amid the complexity of concepts about desertification. In this domain, it is essential to distinguish some concepts that do not correspond to the semantic issue of the term "desertification" but rather to perspectives of methodological approaches for its differentiated qualification and treatment.

Table 4.4 Concepts about desertification according to different authors

Criteria	Concepts	Temporal and spatial scales	Consequences	Authors
Human	Degradation of vegetation cover for the development of cultivated and partial fields	Human; arid and semi-arid environments	Continuous environmental degradation with drying of soils and plant stripping	Abreuvile, (1949)
Human/climatic	Desertification is caused by human action or climate change	Human/geological; arid and semi-arid	Diffusion of conditions of desert environments in arid or semi-arid regions	Rap (1974)
Human	Decrease and destruction of earth's biological potential	Present tense; arid, semi-arid, and dry sub-humid	It leads to a desert (understood as climatic dryness)	Kenya (1977)
Human	Degradation of various types of vegetation and areas with average rainfall between 50 and 300 mm/year	From the recent past to the present time; Margin of deserts and humid forests	Degradation of various types of vegetation	Le Houérou (1977)
Human	Desertification in semi-arid and dry sub-humid geotopes in their ecotones, and the problem of savannization in intertropical Brazil	Human	Degradation of ecological tissues and decrease in biological exploitation factors	Ab'Saber (1977)

(continued)

Table 4.4 (continued)

Criteria	Concepts	Temporal and spatial scales	Consequences	Authors
Human	Conditioned by the instability of the ecological balance, resulting from the rainfall regime with low and irregular rates, shallow soils with low water retention capacity, ample photoperiod, and dehydrating dry and hot winds It presents bioindicators that show the variation of climate elements and socioeconomic conditions. It is a process of fragility of dryland ecosystems that, due to human pressure, or sometimes by native fauna, lose their productive and recovery capacity	Human	Degraded areas with accentuated weakening under conditions of irreversibility of vegetation and soil cover, presenting themselves as small implanted deserts, with the formation of desertification nuclei	Vasconcelos Sobrinho (1978)
Human/climatic	Environmental deterioration process; climatic changes and vicissitudes and inadequate land use. With climate change (global or regional), the process tends to desertification. In the absence of such a change and human action being decisive, it tends to semi-desertification	Human; sub-humid and semi-arid domains in the warm regions of the world	Progressive rainfall deficiency, macro-regional climate changes, and gradual transformation of forested areas into desert areas expressed in the drying of the soils	Nimer (1988)

(continued)

Table 4.4 (continued)

Criteria	Concepts	Temporal and spatial scales	Consequences	Authors
Human/climatic	Population density ratio in 1980 (one person for 10 ha of dry area) promotes symptoms of saarization	Human/geological; arid and semi-arid environments	Progressive disruption of the balance between vegetative associations, the water cycle, agricultural production, the economy, and the social aspect. The lack of conservationist plans in the Portuguese land intensifies the destruction and increases the desert. Immediately, this scenario requires a lot of knowledge, work, and cooperation to avoid the formation of a desert	Duke (1980)
Human	Depletion of terrestrial ecosystems as a result of human activity	Human/geological arid and semi-arid regions	Reduction of agricultural productivity, biomass, micro and macro fauna and flora, soil degradation, and increased risk of degradation for cultivated lands	Dragne (1987)
Anthropic/climatic	It occurs by human or natural action and always culminates in the formation of deserts	Geological and socioeconomic	Aridization and decrease in productive activity. Destruction of the biosphere's potential culminates in a desert	Rozanov and Zoon (1982)
Human	Degradation of "ecological capital"	Human; Sahelian and Sahelosudanese zone	Decreased quantity and productive capacity of resources in water, soil, vegetation, and fauna	Rochette (1989)
Human/climatic	Creation of desert-like conditions, human (ecological) desertification; water deficiency in the natural system, natural (climatic) desertification. Or even both processes simultaneously	Human or geological; sub-humid and semi-arid domains in the warm regions of the world	Natural: deregulation of the hydrological cycle, reduction of precipitation and relative humidity of the air, quaternary oscillations of the tropical arid belts; human: progressive loss of ecosystem productivity, erosion of the surface mantle, elevation of albedo, invasion of sands, desertification points in Paraná and Rio Grande do Sul	Conti (2003)

(continued)

Table 4.4 (continued)

Criteria	Concepts	Temporal and spatial scales	Consequences	Authors
Human/climatic	Questioning about the reduction of annual rainfall totals from social activities	Geological/human; arid and semi-arid lands	Biomass change, accelerated soil erosion, excess monoculture, overgrazing, deforestation, and salinization	Goudie (1990)
Climate and, above all, human	Degradation of drylands resulting, above all, from anthropogenetic impacts	Human; arid, semi-arid, and sub-arid lands	Degradation of bioproductive capacity	UNEP (1995)
Human/climatic	Artificially caused global climate change affects mainly semi-arid regions, which are already the poorest on the planet and have ecosystems of high environmental vulnerability	Human; sub-humid and semi-arid domains in the warm regions of the world	Growing loss of productive capacity of ecosystems; possible irreversibility of the process; major and drastic social consequences	ICID (1992)
Human	Destruction of the biological potential of arid and semi-arid lands	Present tense; arid, semi-arid, and dry sub-humid	Deterioration of life, interface in the fragile relationship man/climate/soil/vegetation	Rio, ECO (1992)
Human/climatic	Consideration of Unep's aridity index and use of desertification indicators	Human	Areas susceptible to desertification occur in the Brazilian Northeast	Valdemar Rodrigues et al. (1992)
Human	Human interference in the physical environment causes disruptions in geoecological dynamics and land degradation can be irreversible	Human: recent past to today	It can occur in virtually any climate zone	Dregne (1987)
Human/climatic	There are generalizations, confused and mistaken interpretations in the concept of desertification	Human or geological	Deep and large environmental and socioeconomic damages	Rubio (1995a, b)

(continued)

Table 4.4 (continued)

Criteria	Concepts	Temporal and spatial scales	Consequences	Authors
Human	Irreversibility	Temporal human (25 years); arid, semi-arid, and dry sub-humid areas	Destruction of the biological potential of land and the ability to support populations	Mainguet (1995a, b)
Human/climatic	Difficult reversibility; process accelerated in drylands by climate change and greenhouse gases	Geological and human, associated	Drylands advancement; frequent extreme events	Pereira and do Nascimento (2016, 2020), do Nascimento (2023) and IPPC (2021)
Human/climatic	Land use and climate change accelerate desertification	Human rather than geological	Degradation of natural resources; increase in droughts, changes in land use	Bezerra (2020), Vieira et al. (2020) and Birth (2020)

Source do Nascimento (2013)

Furthermore, the repercussions of the United Nations Conference on Environment and Development (UNCTAD), one of the most extensive global ecopolitical meetings in recent years, held in Johannesburg, South Africa, in 2002, did not bring due importance to the discussion of desertification. Rio+20 was equally disappointing.

References

Ab'saber (1977) Aziz Nacib. A problemática da desertificação e da savanização no Brasil. In: Geomorfologia, n° 53. USP, São Paulo, p 20

Bezerra FGS (2020) Analysis of areas undergoing desertification, using EVI2 multi-temporal data based on MODIS imagery as indicator. Ecol Ind 117:106579

Brasil (2004a) Caderno de debates. Agenda 21 e Sustentabilidade: Agenda 21, o Semi-árido e a luta contra a desertificação. N° 6. Brasília: MMA, 15p. In: Recursos Hídricos: conjunto de normas legais, 3ª edn, Brasília, Ministério do Meio Ambiente/Sec. dos Recursos Hídricos, pp 149–158

Brasil/Ministério do Meio Ambiente (MMA) (2004b) Programa de Ação Nacional de Combate à Desertificação e Mitigação dos Efeitos da Seca, PAN-BRASIL. Edição Comemorativa dos 10 anos da Convenção das Nações Unidades de Combate à Desertificação e Mitigação dos Efeitos da Seca—CCD. MMA, Brasília, 225p

Castro FC, Araujo JF, Santos AM (2019) Susceptibility to soil salinization in the quilombola community of Cupira—Santa Maria da Boa Vista—Pernambuco—Brazil. CATENA 179:175–183

Conselho Nacional da Reversa da Biosfera da Caatinga (Brasil)—CNRBC (2004) Cenários para o Bioma Caatinga. Secretaria de Ciências, Tecnologia e Meio Ambiente. SECTMA, Recife, 283p

Conti JB (2003) A Desertificação como forma de degradação ambiental no Brasil. In: da Ribeiro W (Org) Patrimônio ambiental do Brasil. Edusp: Imprensa Oficial do Estado de São Paulo, São Paulo, pp 167–190

da Vieira RMSP (2020) Characterizing spatio-temporal patterns of social vulnerability to droughts, degradation and desertification in the Brazilian northeast. Environ Sustain Ind 5:100016–100019

do Nascimento FR (2013) O Fenômeno da Desertificação. 1st edn. Goiânia: EDITORA UFG, 1:240

do Nascimento FR (2021) Desertificación—Diccionario de desarrollo regional y cuestiones conexas. Diccionario de Desarrollo Regional y Cuestiones Conexas, vol 2, 2nd edn. Editora Conceito, Uruguaiana, pp 245–249

Dregne HE (1987) Envergadura y difusión del proceso de desertificación. In: Programa de las Naciones Unidas para el Médio Ambiente (PNUMA): Comision de la URSS de los Asuntos de PNUMA. Colonizacion de los territórios áridos y lucha contra la desertification: enfoque integral. Moscu, Centro de los Proyectos Internacionales - GKNT, pp 10–17 (Capitulo I)

Drew D (1986) Processos interativos homem-meio ambiente. (Trad. dos Santos JA, Bastos S). Difel, São Paulo, 206p

Here FK et al (1992) Desertificação: uma visão global. In: Desertificação causas e consequências. (Trad. Port) Lisboa, Fundação Caloute Gulbenkina, pp 12–108

IPPC (2021) The Intergonvernmental panel on climate change. Geneva. Cite on http://www.ipcc.ch, 02 Jan 2021

Mainguet M (1995a) La désertification expression de la décadence? In: L'Homme et la sécheresse. Masson, Paris, pp 285–296

Mainguet M (1995b) Les notions d'aridité et de sécheresse dans les écosystèmes secs. In: L'Homme et la sécheresse. Masson, Paris, pp 27–50

Nimer E (1988) Climatologia do Nordeste. Desertificação: mito ou realidade. Rio de Janeiro: Ver. Brasileira de Geografia. IBGE 50(1):7–39

Pereira JI, do Nascimento FR (2016) Susceptibility to desertification in Chicualacuala, Republic of Mozambique. Int J Geosci 07:229–237

Pereira JI, do Nascimento FR (2020) Focusing on the susceptibility to desertification in Chicualacuala, Republic of Mozambique. In: Yu H (Org) Focusing on the susceptibility to desertification in Chicualacuala, Republic of Mozambique, vol 5, 1st edn. Internacionational, Inglaterra, pp 18–29

Pernambuco (2001) Política Estadual de Controle da Desertificação. Secretaria de Ciências, Tecnologia e Meio Ambiente, Recife, 364p

Redesert (2003) Rede de Informações e Documentação em Desertificação. O que é desertificação. Brasília, Governo Federal, pp 2–8

Rozanov BG, Zoon S (1982) The definition, diagnosis and assessment of desertification. In: relation to experience In: the USSR. Dertification control, 7:13–17

Rubio JL (1995a) Desertification: evolution of a concept. In: Seminário Desertificación y Cambio Climático. Centro de Investigaciones sobre Desertificación – CIDE/Universidad internacional Menendez Pelayo (UIMP), C.S.I.C – Valencia, p 9

Rubio JL (1995b) Definiciones. Marco Conceptual. In: Seminário Desertificación y Cambio Climático. Centro de Investigaciones sobre Desertificación – CIDE/Universidad Internacional Menendez Pelayo (UIMP), C.S.I.C – Valencia, p 46

Rubio JL, Bochet E (1998) Desertification indicators an diagnosis criteria for desertification risk assessment in Europe. In: International symposium and workshop – combating desertification: connecting science with community action, J Arid Environ, vol 39, article n° ae 980402. Academic Press Limited. Tucson, Arizona, USA, pp 113–120

Suertegaray DM (2003) Desertificação – recuperação e desenvolvimento sustentável. In: Guerra AJT, Cunha SB da (orgs) Geomorfologia e Meio Ambiente. Rio de Janeiro, Bertrand Brasil, pp 249–290

United Nations Environment Programme (UNEP) (1995) News of interest. In: Desertification control bulletin: a bullet in: of world events in the control of desertification, restoration of degraded lands an reforestation, vol 27. United Nations Environment Programme (UNEP), pp 93–96

Verdum R et al (2002a) Desertificação: questionando as bases conceituais, escalas de análise e conseqüências. In: GEOgraphia, Revista da Pós-Graduação em Geografia do Departamento de Geografia da Universidade Federal Fluminense. Ano 3, n° 6, Niterói, UFF/EGG, pp 119–132

Verdum R et al (2002b) Tratados Internacionais e Implicações locais: a Desertificação. In: GEOgraphia, Revista da Pós-Graduação em Geografia do Departamento de Geografia da Universidade Federal Fluminense. Ano 6, n° 11. Niterói, UFF/EGG, pp 79–88

Verdum R (2004) Tratados internacionais e implicações locais: a desertificação. GEOgraphia (UFF), 11:79–88

Zonn IS, Orlovski NS (1987) Factores antropogênicos de la desertificación. In: Programa de las Naciones Unidas para el Médio Ambiente (PNUMA): Comision de la URSS de los Asuntos de PNUMA. Colonizacion de los territórios áridos y lucha contra la desertification: enfoque integral. Centro de los Proyectos Internacionales—GKNT, Moscu, pp 17–24 (Capitulo II)

Chapter 5
Deserts, Desertification and Environmental Degradation in the World

Abstract *Literature and epistemological reviews of the construction of the* desertification concept demonstrate that the concept was forged to portray the problems of drying soils and expansion of deserts in the Sahel, generalizing similar occurrences to desertification in the rest of the planet. This context demands regional and local studies in search of better knowledge of the environmental problems of areas affected by this problem. For the bases of this question, a more and better understanding of the concepts and main characteristics of deserts and the desertification phenomenon, desert classifications, gauging the areas affected by desertification in the world, and associated causes, are indispensable. From this, recognize the extent of the drylands of the globe and their geographic distribution, in absolute and relative terms, for greater accuracy of scientific research. Based on this, the international experiences and impacts resulting from desertification at a planetary, regional, and local level are better recorded.

Keywords Deserts · Desertification · Drylands · World · Environmental problems

5.1 Classifications and Types of Deserts

Whether by natural or socioeconomic influence, degradation/desertification has as its tonic the degradation and drying of the soil, which can culminate in environmental exhaustion. As highlighted by UNEP, desertification is "environmental drying, produced by the impact resulting from human activities that cause the degradation of previously productive lands" (Nimer 1988, p. 16).

Without denying the diversity of approaches and generic understandings about desertification, in the strictly climatological scope, desertification would only occur in a considerably large area if a humid or semi-arid macroclimate was transformed into a desert or semi-desert macroclimate. A desert origin links to a profound change in the thermodynamic balance of the atmosphere, which totally or partially surrounds the planet (Nimer 1988; Batchelor and Wallace 1995).

© The Author(s), under exclusive license to Springer Nature Switzerland AG 2023
F. Rodrigues do Nascimento, *Global Environmental Changes, Desertification and Sustainability*, SpringerBriefs in Latin American Studies,
https://doi.org/10.1007/978-3-031-32947-0_5

Desertification as an increasing dryness of the physical environment can result from changes in the regional climate, inadequate land use, or both cases simultaneously. Nature degradation alone, however, cannot trigger a drying typical of the phenomena of desert formation or similar to these.

Regardless of the desert, it is basically determined by the regional macroclimate, determined by the constant high atmospheric pressure cell over the region. It is, therefore, a natural phenomenon, independent of human intervention. This perspective differs from the official concept of desertification. In this context, Nimer (1988) defines desertification as a set of phenomena that lead certain areas to become deserts or resemble them. They result from climate change under natural and socioeconomic influences—pressure on fragile ecosystems, among which the outskirts of deserts (or transitional areas) are at greater risk due to the weak environmental balance.

Cavalcanti (2003) and do Nascimento (2013) ensure that the word "desertification" has been leading to misinterpretations, mainly when associated with the process of creating deserts. It should be noted that deserts are specific ecosystems with their genesis and dynamics. Strictly speaking, the desert concept cannot be confused with that of desertification, nor should the processes of environmental degradation in drylands be equated with the emergence of a desert biome. It is necessary to conceptualize desert and list the differences between it and desertification, as well as apprehend each process's respective causes and consequences.

Even though the words "desert" and "desertification" have the same etiology, they relate to different phenomena. Strictly, "desertification" comes from the Latin *deserta facere*, something like "to make or manufacture a desert". Regardless of variations on the word "desert", all official UN languages have a common idea—that deserts are strange, lifeless, homeless, or even empty places. Nimer (1980) exemplifies that in Chinese, desert means little water or strange thing; in Russian and Arabic, desert and void have the same origin; in Portuguese, Spanish, French, and English, the Latin root of the word *desertus* means abandonment, depopulated.

According to Ricklefs (1996) and do Nascimento (2013), deserts are formed by climatic conditions. Dry air masses reach the surface in subtropical latitudes, spreading to the North and South after condensing in intertropical areas, with a decrease in their ability to evaporate and retain water, increasing as they descend and are heated. This time, they begin to extract steam from the earth, causing climatic aridity centered on latitudes close to 30° north and south of the equator, ranges in which the great hot deserts develop: Kalahari, Namibia, Arab and Sahara, in Africa; Sonoran, Chihuahuan, Mohave, North America; the Atacama in South America; and the Australian desert.

Exceptions to this dynamic are due to orographic barriers (mountains, mountain ranges, escarpments, slopes, etc.), which cause cooling and loss of water vapor, causing precipitation on the windward slope, where the humid winds blow. In the opposite position is the slope protected by the wind—called leeward—with air that descends toward the continent and enters it, drying it and depriving it of moisture to create arid environments called rain shadows. The upward movement of the air, influenced by the mountains, cools it, and its vapor is condensed, causing precipitation and providing more humidity on this slope, which causes climatic rigors conditioning the formation of natural deserts in rain shadows, with large mountainous extensions. This case is the situation of the Great American Basin (USA) deserts and the Asian Gobi.

Except for Southeast Asia and Northeast Australia, where the monsoon system disrupts atmospheric circulation, latitudes between 20° and 30° are generally vulnerable to desert conditions, and natural deserts occur in subtropical areas of high atmospheric pressure, between 15° and 25° north and south latitude (Zonn and Orlovski 1987).

In continental areas with atmospheric stability, located between latitudes 20° and 35°, in which water replenishment is insufficient to restore water balance, there are possibilities for the emergence of deserts. The tropics lie at 23° 27′ north and south latitude. These are areas of high pressure where the winds blow toward lower latitudes, that is, toward the equator. Thus, the air descends from the top of the atmosphere, becoming hotter and drier.

It is possible to classify deserts based on these climatic, morphological, biological, geographic, and paleogeographic aspects (landfills, mountain framing and sinking of their bases, etc.). The proposals for classifying these biomes in their morphoclimatic zones are shown in Table 5.1. A hot and dry desert, strictly speaking, is a natural region where it rains up to 250 mm/year or remains without rain for 50 years or more. They can be of two types: *ergs*, large sheets of sand, or *hamadás*, "deserts of stones", immense deposits of dismantled rocks, especially sandstones, which can decompose quickly, facilitating the infiltration of water. Both landscapes generally harbor few forms of life, with almost no plant formations.

These desert regions have the highest daily temperature ranges on the entire planet. The rains are concentrated over a month or even in a few days. In hot deserts, the water balance is balanced only in the short period of rain, remaining for ten months or more with a marked deficit concerning the biological need for rainwater. In them, the actual evapotranspiration is less than the potential evaporation during most of the year.

The general climatic characteristics display droughts, thermal amplitude, and air agitation. This one has expressive dynamics and extreme energy, with displacement and movement of fine dust, which can reach 2 km in height and travel hundreds of kilometers. The *loess* of Northern China, areas of accumulation of fine deposits transported by the wind action, located on the margins of ancient areas of glaciers or the periphery of deserts, come from Upper Central Asia, according to Pouquet (1962). The *loess* present a capping of up to 100 m, with the surface of the ravines measuring about 26 thousand km^2 and the total surface of the soils 600 thousand km^2 (Dregne 1987). These dynamics happened due to anticyclonic and digressional air masses of planetary order and thermometric variations that influence barometric pressures.

Loess also *occurs* in southern Russia, the central part of the USA, and eastern Argentina. In general terms, high temperatures and wide dinoturnal and annual amplitudes are recorded in some cases.

Table 5.1 Classifications of deserts

Proposals	Typologies
1942—E. F. Gautier, Emanuel De Martone, Aufrère and Kachkara classify the world's deserts according to geographic aspects	*Hot deserts*: Absolute or classic types—Central and Atlantic Sahara, Atacama, Libya, and perhaps central Australia; deserts of the Chilean, Peruvian and South African coasts—Namibia—associated with cold currents Attenuated types—Greater Chaco, Calaari, Mauritania, Western Australia, Sahara mountain ranges *Cold deserts*: Temperate deserts, whose only representative is the attenuated type, such as Northern Patagonia, Siankiang, Middle Asia, Iranian Plateaus, Syrian Desert, and the Great American Basin Cold deserts, mainly at altitude, with sufficient examples in Gobi and Tibet
1947—De Martone, climate classification of the world's desert regions	*Rigorous deserts*: Saharan and Chilean types, from South Africa and the island of Madagascar, for example *Attenuated deserts*: harsh winter and cold high altitude deserts of: North Central Asia, Patagonia, and West and Southeast USA *Deserts*: Steppe degradation of temperate and Mediterranean climates, in which its genesis is not correlated with Nimer's theory of semi-desertification (1988): American Midwest, Southern Brazil and Argentine Pampas, margins of the attenuated deserts of Asia • Deserts under the influence of Monsoon (Sahara): enclave
1990s—Steele, geographic and climate classification	*Continental deserts*: Continental deserts are those that form in the center of large tracts of land and are located at higher latitudes. They can be cold and have freezing winters. They are dry because the oceanic humid winds arrive dry in the center of the continents *Deserts that formed along coastal regions*: Cold ocean currents encountered a very warm continent. The heat removes moisture from the clouds formed over the ocean, causing permanent soil dryness. So are the Namibian and Atacama deserts. These two deserts may also be of leeward origin. For example, the deserts of Death Valley, on the border between California and the state of Nevada, in the western US, located in the northeast of the Majave Desert and west of the Amargosa Desert *Frozen deserts*: Frozen deserts receive very little rain and, although they hold large volumes of water, they are frozen

Source Pouquet (1962), Rubio (1995a), Steele (1998), Conti (2003) and do Nascimento (2013)

5.2   Deserts, Desertification and Drylands

We can also assure that the rainfall regime in deserts is very irregular, both in form and frequency, with consecutive years of insignificant precipitation (as in the Australian desert) or even non-existent (in the Saharan core). However, the majority of rainfalls are torrential, causing *sheet-flood* flow. However, there are also manifestations of precipitation in the form of weak and lasting rains.

As a bioecological reflection of the hydrological framework, the deserts present only tenuous vegetation of herbaceous predominance, with the edaphic conditions of the substrate minimally favorable to the phenology of more complex plants.

Vegetation can prevent wilting and remain dormant for long periods, increasing the economic efficiency of transpiration. Thus, the proportion of dry matter produced concerning water transpired is higher than in plants that are not typical of deserts (Nimer 1980).

Natural population control mechanisms are efficient to avoid competition. As an example, the vegetation spacing with exposure of large expanses of bare soil and phytocenosis of little diversity of species, in which the dominant species are relatively abundant, prove the existence of this type of ecological relationship.

Demographics for the 2000s showed that about a third of the world's population inhabits arid or semi-arid areas. The dry tropics (< 600 mm/year) belong to the climatic zone corresponding to about 40% of the earth's surface, comprising a large part of poor, agricultural countries with low per capita income (do Nascimento 2016; Ayoade 2002). For the synthetic effect of comparison, Table 5.2 demonstrates some aspects related to deserts and desertification.

Table 5.2 Concepts and main characteristics of desert and desertification phenomenon

Concepts	Main features
Desert: any area where rainfall is equal to or less than 250 mm/year	Aridity conditioned by: (a) semi-permanence of high pressures of tropical or subtropical anticyclone in place; (b) leeward positioning of an orographic system, with an altitude sufficient to completely or partially block the mechanisms of rain; (c) high altitudes, above the water vapor condensation level; (d) proximity to ocean currents of cold water, drastically reducing evaporation and air humidity and, by extension, rainfall in the surrounding areas. It refers to the idea of climate type, with adaptation of the natural system, identity, and defined spatial limits. It is often an arid region, with potential evaporation greater than annual precipitation and poor biosphere development. Rainfall is irregular yearly. Shallow soils, with low edaphoclimatic potential, tend to concentrate salts. Sparse vegetation, with adapted xerophytes and fauna. Erosive processes are driven mainly by winds (simum, harmattan, etc.). Rare and torrential rains, with turbulence. Generally, the central area has severe aridity and a less dry periphery and transition to sub-humid areas. Life in deserts is regulated by the availability of water, which influences the quality of soils and the presence or abundance of plants, and these govern the presence of animals
Desertification: as a substantiation of the word desert, it indicates a phenomenon of continuous dynamics, with its environmental evolution heading toward a desert or similar to it	It has a process and dynamism associated with prolonged dry periods, on the order of decades. It is a phenomenon of natural imbalance, mainly characterized by an increasing dryness of its environment, determined by changes in the regional climate and/or soils; or both, concurrently. It should be noted that the degradation of the environment is incapable of changing the regional climate to the point of causing a drying characteristic of a genetic process of natural formation of a physical-ecological desert

Source Pouquet (1962), Nimer (1988), Conti (1994) and do Nascimento (2013)

In fact, for the last 2 thousand years, desertification has existed due to human activities. During a much more extended period, deserts, in turn, appeared due to climatic fluctuations, as in the example of the Sahara (Dregne 1987). Aware of these issues, desertification is best understood as the degradation of the physical environment in drylands. It is a process enclave that can vary, with regeneration periods, being economically reversible during its development. The desertification origins and consequences vary regarding ecosystem variables and land-use history. It is a slow process, which may occur not in years but decades of observation (Stiles 1995). On the other hand, the containment of a natural desert, that is, physical-ecological, would be impracticable.

5.2.1 Environmental Degradation, Desertification in the World and Environmental Impacts

Three hundred years ago, the areas covered by forests corresponded to almost two-thirds of our planet. Deforestation was so intensive that vegetation formations today cover less than a third of the continents. Since 1980, it is estimated that arid zones in the world have grown by around 15% (Steele 1998).

In fact, since the birth of macroeconomics in the nineteenth century, with the Industrial Revolution that subjugated low latitude countries to suppliers of raw materials and primary products, associated with the expansion of a materialistic culture and production activities on a large scale, a new relationship between human society and nature, dilapidating much more than preserving (Conti 2002).

Without forgetting, still according to Conti, the elements of the physical environment in the tropics almost always have weaknesses in their balance; any form of environmental destabilization can cause or accentuate weathering, soil leaching, slope instability, and generalized degradation.

Desertification affects 33% of the earth's surface, where more than 2.6 billion people live, or 42% of the world's population. Its effects are exacerbated in the affected sub-Saharan region, with more than 200 million inhabitants, representing 20–50% of the degraded lands. In Asia and Latin America, for example, land degradation is severe (Brasil 2004).

According to the UN, between 1983 and 2013, 20% of the planet's land was degraded, and 2 billion hectares worldwide deteriorated. However, most land can be recovered (UN 2021).

The planet has a surface area of 510,110,934 km^2, of which the continents occupy about 140 million. Drylands, according to the United Nations Environment Program—(UNEP) 1995, corresponding to about 47.2% of all emerged lands in the world, distributed in the following climatic ecozones: hyperarid—7.5%; arid—12.1%; semi-arid—17.7%; and dry sub-humid—9.9% (Batchelor and Wallace 1995). Semi-arid lands exist in all latitudes, but the most problematic areas, concerning the loss of the bioproductive capacity of the environment, are located in tropical developing countries due to the production of space, historically and territorially degrading.

These areas are home to indigent populations with a disastrous quality of life—low levels of income, low levels of technology and education, and protein intake below acceptable levels by the World Health Organization (WHO); however, that process of environmental deterioration has a specific evolution in each place, based on its dynamics (Cavalcanti 2003; ONU 2021, 2022).

The dry areas can be expressed in millions of hectares, showing the total land on the planet and its respective climatic areas, approximately as shown in Table 5.3, considering the indices already listed. It is worth remembering that hyperarid lands, considered unproductive—except for those with meager yields—are not considered in the measurement of desertification, as is the Antarctic continent, which, despite having 13,340,000 km^2, does not present drylands. This table shows that Africa has the most expressive area with dry climatic ecozones, 19,590,000 km^2, followed by Asia, with 100,000 km^2 less, or 31.8% of the total. Of the global total of 61,500,000 km^2 of dryland, South America boasts 5,430,000 km^2 or 8% of the total. In particular, South America has 260,000 km^2 and 45,000 km^2 of hyper-arid and arid lands, represented by the Atacama Desert, and 2,650,000 km^2 of semi-arid, in which the Brazilian Northeast is inserted. Last in this *ranking* comes Europe, 3,000,000 km^2, or 5% of the total.

The VI Meeting of America and the Caribbean, held in San Salvador in October 2000, with the participation of 33 countries, has already indicated that approximately 170 million hectares have been degraded in South America due to deforestation and overgrazing. In the Caribbean, accelerated and poorly planned urbanization has resulted in land losses for agriculture, unprotected watersheds, and a decline in biodiversity conservation, contributing to the advance of desertification. In addition, 63 million hectares were compromised in Mexico and Central America.

Table 5.3 Drylands on the globe and their geographic distribution, in absolute (millions of km^2) and relative (%) terms

Typologies climatic	Africa	Asia	Australia	Europa	America from north	America South	Global total drylands
Hyperarid	6.720	2.770	0	0	30.000	260	9.780
Arid	5.040	6.260	3.030	110	820	450	15.710
Semi-arid	5.140	6.930	3.090	1.050	4.190	2.650	23.050
Sub-humid dry	2.690	3.530	510	1.840	2.320	2.070	12.960
Total area	19.590	19.490	6.630	3.000	7.360	5.430	136.224
Relative area (%)	32	31.8	11	5	12	8	100
Total area of continent	30.335	43.508	8.923	10.498	25.498	17.611	61.500

Source Stiles (1995)

In Latin America, the most problematic countries are Argentina, Bolivia, Brazil, Chile, Ecuador, and Peru. They have sought to unite to face desertification, not only with public policies but also through a full articulation in favor of the issue. In particular, the Southern Cone countries—Argentina, Brazil, Paraguay, and Uruguay—seek to articulate themselves to combat desertification. The issue of water resources, in which hydrographic basins are the units of study, is the most relevant, which helps in the integration between these countries. Furthermore, it should be noted that the synergy between the conventions that deal with the desertification theme favors the integration of these nations. At the regional level, the Grande American Chaco stands out, involving Argentine, Bolivian, and Paraguayan areas, with one million square kilometers and a population of over four million, highlighting this problem in a sub-regional action program.

In Argentina, Escobar (1997) alerted that 60 million hectares suffer different types of erosion; annually, 650,000 ha are degraded, and about 26% of the population lives in semi-arid or arid areas. With this situation, there are severe economic losses in the province of Chubut, with the loss of about hundreds of permanent jobs and thousands of temporary work fronts per year.

In Europe, the most affected countries are Greece, Italy, Portugal, France, Albania, Turkey, and Bulgaria—see Fig. 4.1. However, according to Rubio (1995b), the most worrying is Spain, where the problem affects 60% of the territory.

The main strip of desertification on the planet is located in West Africa, bordering the southern tip of the Sahara, covering 37 countries and extending from the Atlantic Ocean to the Indian Ocean, with the Sahara-Savanna ecotone. It brings together countries such as Senegal, in the Atlantic Ocean, Ethiopia, and Somalia, in the Indian Ocean, and others such as Mali, Niger, and Burkina Faso, to name some of the most affected by the drying of the soil. The eight most affected countries are Angola, Burkina Faso, Chad, Ghana, Mauritania, Mozambique (Pereira and do Nascimento 2020), Niger, and Zimbabwe.

In the mid-1980s, this region again suffered a terrible drought, the consequences of which further aggravated the ills of desertification. In Ethiopia alone, the droughts affected around seven million people, which caused many populations to migrate.

In Asia, the hardest-hit countries are China and India. It is worth remembering that, although climatic contingencies, as a rule, are not so relevant in developed countries, their economies and environment suffer losses in semi-arid regions, as, for example, in the great American and Canadian plains, as well as in central Australia.

Considering the internationally accepted typology of areas susceptible to desertification, highlighted by UNEP, and the information gathered in Table 5.4, we can say that the areas affected by desertification and their related causes can be arranged according to the table below. Unfortunately, most of these areas coincide with underdeveloped countries' most significant pockets of poverty. Compounding the problem, desertification, until the end of the last decade, affected around six million hectares ($60,000 \text{ km}^2$) per year, in which overgrazing, soil salinization by irrigation, and intensive use processes without agricultural management worsened the situation (do Nascimento 2013).

Table 5.4 Areas affected by desertification in the world and associated causes

Types	km²	% of total dry areas
1. Areas degraded by irrigation	430,000	0.8
2. Areas degraded by rainfed agriculture	2,160,000	4.1
3. Areas degraded by livestock	7,570,000	14.6
4. Dry areas with soil degradation (1 + 2 + 3)	10,160.00	19.5
5. Degradation of grazing lands	25,760,000	50
6. Total degraded dry areas (4 + 5)	35,920,000	69

Source Redeser—Information and Documentation Network on Desertification (2003) in do Nascimento (2013)

Population data (one-sixth of the world population) and the area affected by desertification are of exaggerated magnitude, but it is the dimension that subsidizes the studies and serves as a basis for questioning the problem (Verdum 2004).

As for soil salinization, Furriela (2002), Ayers and Westcot (1991) state that this type of impact causes plant death due to the compromise of structure and toxicity, increasing the vulnerability of water and wind erosion; on a global scale, it affects about 20% of irrigated soils. Extrapolating these numbers from 25 to 30%, Brazil (2004) highlights that 274 million irrigated hectares in the world have problems with salinization and soil saturation, among other problems, due to inadequate management of irrigation and drainage in land use. There is no consensus, however, as to precise numbers on the issue, as its concepts, perspectives, approaches, and indicators are multiple and varied.

Bringing more details to this issue, Dregne (1987) and Castro and Santos (2020) say that the main factor of pasture lands desertification is vegetation destruction; for lands with rainfed agriculture, soil erosion is the central degradation; in irrigated agriculture, salinization is one of the great villains. Erosion, soil crusting, and soil compaction are secondary desertification factors in pasture areas. In addition to these aspects, in rainfed agriculture, among the secondary factors, stands out the fertility decrease. In addition to the loss of nutrients, the sealing of the land increases flooding, and the capture of carbon and other greenhouse gases interfering with climate change are consequences of poorly managed and degraded soils.

FAO's critical 2015 report, Status of the World's Soil Resources, claims that 33% of the world's soils are degraded. The main problems are erosion, salinization, compaction, acidification, and contamination.

Among saline (412 million ha) and sodic (618 million ha) soils in the world, there are 1030 million ha affected. The most problematic continent is Africa, with 122.9 million ha salinized and 86.7 million soda (UNEP 1992 in FAO 2015).

From an economic-ecology perspective, soil erosion, salinization, and desertification are neglected aspects of development processes and directly affect the economy's sustainability. Conti (2003) believes that the advance of desertification is directly linked to the degree of development and highlights UNEP estimates that 86% of productive lands in Africa are affected by this phenomenon. In this proportion are some of the poorest African countries; in Australia, a rich country, only 22% of productive land is affected by desertification. As about 97% of our food comes from emerged lands, it is worrying that the arable soils on the planet are already at high levels of degradation.

Conti (2002) considers desertification an irreversible phenomenon on a socioeconomic time scale for humanity, when 2.3 billion people (or 30% of the global population) do not have access to adequate food throughout 2020– since the beginning of the COVID-19 Pandemic, as several international organizations point out: Food and Agriculture Organization of the United Nations, The International Fund for Agricultural Development (IFAD), United Nations International Children's Emergency Fund (UNICEF), and World Food Program (WFP) (2021).

International events on climate change, such as COP26, essential reports, and works, such as the Great Green Wall/GGW of (UNCCD) (2020), highlight the need to scale up the United Nations Decade of Action for Nutrition (2015–2025) in direct consonance and indirectly with the Sustainable Development Goals 2030—SDGs (FAO, IFAD, UNICEF, WFP and WHO 2021; FAO 2022).

Furthermore, there is a need for a better understanding of the complex relationship between desertification and climate change—which causes degradation and famine. The origin of environmental degradation may link to possible global climate changes expressed, above all, in a growing rainfall deficiency—and to human action or both factors simultaneously.

References

Ayers RS, Westcot DW (1991) A qualidade da água na agricultura. Estudos FAO—Irrigação e Drenagem. UFPB, Campina Grande, 218p

Ayoade JO (2002) Introdução à climatologia para os Trópicos. 8ª edn. Bertrand Brasil, Rio de Janeiro, 332p

Batchelor CH, Wallace JS (1995) Hydrological process, dryland degradation and integrated catchment resource management. In: Desertification control bulletin: a bulletin: of world events in the control of desertification, restoration of degraded lands an reforestation, vol 27. United Nations Environment Programme (UNEP), pp 27–34

Brasil/Ministério do Meio Ambiente (MMA) (2004) Programa de Ação Nacional de Combate à Desertificação e Mitigação dos Efeitos da Seca, PAN-BRASIL. Edição Comemorativa dos 10 anos da Convenção das Nações Unidades de Combate à Desertificação e Mitigação dos Efeitos da Seca—CCD. MMA, Brasília, 225p

Castro FC, Santos AM (2020) Salinity of the soil and the risk of desertification in the semiarid region, vol 19. Mercator, Fortaleza, pp 1–13

Cavalcanti E (2003) Para Compreender a desertificação: uma abordagem didática e integrada. Governo do Estado de Pernambuco, Recife, 56p

Conti JB (1994) O conceito de desertificação. In: Anais do 5 Congresso Brasileiro de Geógrafos: Velho mundo—novas fronteiras: perspectivas da Geografia Brasileira, vol 1. AGB, Curitiba, PR, pp 366–368

Conti JB (2002) As relações Sociedade/Natureza os Impactos da Desertificação nos Trópicos. Cadernos Geográficos. N° 04. Florianópolis: Ed. UFS C

Conti JB (2003) A Desertificação como forma de degradação ambiental no Brasil. In: da Ribeiro W (org) Patrimônio ambiental do Brasil. Edusp: Imprensa Oficial do Estado de São Paulo, São Paulo, pp 167–190

do Nascimento FR (2013) O Fenômeno da Desertificação. Cegraf, UFGO, Goiânia, 240p

do Nascimento FR (2016) Os semiáridos e a desertificação no Brasil. REDE—Revista Eletrônica do PRODEMA, Fortaleza 9(2). Disponível em: http://www.revistarede.ufc.br/rede/article/view/312. Acesso em: 24 Jan 2022

Dregne HE (1987) Envergadura y difusión del processo de desertificación. In: Programa de las Naciones Unidas para el Médio Ambiente (PNUMA): Comision de la URSS de los Asuntos de PNUMA. Colonizacion de los territórios áridos y lucha contra la desertification: enfoque integral. Centro de los Proyectos Internacionales—GKNT, Moscu, pp 10–17 (Capitulo I)

Escobar JM (org) (1997) Desertificación em Chubut. Proyecto de Prevención y Control de la Desertificación para el Desarrollo Sustentable de la Patagônia (PRODESAR). Proyecto Argentino Aleman—INTA/GTZ: Instituto Nacional de Tecnologia Agropecuária; Centro Regional Patagônia Sur y Estación Experimental Agropecuária Chubut. Trelew (Argentina), 29p

Food and Agriculture Organization of the United Nations (FAO)/Intergovernmental technical panel on soils (ITPS) (2015) Status of the World's Soil Resources. Main report. Rome, p 650

FAO (2022) Status of the World's Soil Resources. FAO, Roma. Cited on https://www.fao.org/documents/card/en/c/cb4474en. Acessado: 10 Jan 2022

FAO, IFAD, UNICEF, WFP and WHO (2021) The state of food security and nutrition in the world 2021. In: Transforming food systems for food security, improved nutrition and affordable healthy diets for all. FAO, Rome

Furriela RB (2002) Mudanças climáticas globais e biodiversidade. In: Bensusan N (org) Seria melhor mandar ladrilhar? Biodiversidade, como para quê, por quê. Brasília: Ed. da UNB, instituto socioambiental, pp 219–228

Nimer E (1980) Subsídios ao Plano de Ação Mundial para Combater a Desertificação—Programa das Nações Unidades Para o Meio Ambiente (PNUMA). In: Revista do Instituto Brasileiro de Geografia e Estatística (IBGE). IBGE/SPREN 42(2–3):612–637

Nimer E (1988) Desertificação: mito ou realidade. Ver. Brasileira de Geografia. IBGE, Rio de Janeiro 50(1):7–39

ONU (2021) Nações Unidas: degradação de terras impacta 3,2 milhões de pessoas no mundo. Cited on https://news.un.org/pt/story/2018/06/1627442. Acessed:13 Dec 2021

ONU (2022) Terras Áridas. Cited on https://news.un.org/pt/tags/terras-aridas. Acessada em: 24 Jan 2002

Pereira JI, do Nascimento FR (2020) Focusing on the susceptibility to desertification in Chicualacuala, Republic of Mozambique. In: Yu H (Org) Focusing on the susceptibility to Desertification in Chicualacuala, Republic of Mozambique, vol 5, 1st edn. Internacionational, Inglaterra, pp 18–29

Pouquet J (1962) Os desertos. Editora Difusão Europeia do Livro. São Paulo, 127p

Ricklefs RE (1996) A economia da natureza, 3° edn. Guanabara Koogam SA, Rio de Janeiro, pp 53–71

Rubio JL (1995a) Desertification: evaluation of a concept. In: Seminário Desertificación y Cambio Climático. Centro de Investigaciones sobre Desertificación—CIDE/Universidad internacional Menendez Pelayo (UIMP), C.S.I.C—Valencia, 9p

Rubio JL (1995b) Alcance del problema. In: Seminário Desertificación y Cambio Climático. Centro de Investigaciones sobre Desertificación—CIDE/Universidad internacional Menendez Pelayo (UIMP), C.S.I.C, Valencia, pp 14–47

Steele P (1998) Desertos. Editora Scipione Ltda, São Paulo, 32p

Stiles D (1995) Desertification is not a Myth. In: Desertification control bulletin: a bulletin: of world events in the control of desertification, restoration of degraded lands an reforestation, vol 26. United Nations Environment Programme (UNEP), pp 29–36

United Nations Convention to Combat Desertification (UNCCD) (2020) The Great Green Wall implementation status and way ahead to 2030. Bonn, Germany, 45p

Verdum R (2004) Tratados internacionais e implicações locais: a desertificação. GEOgraphia (UFF), 11:79–88

Zonn IS, Orlovski NS (1987) Factores antropogênicos de la desertificación. In: Programa de las Naciones Unidas para el Médio Ambiente (PNUMA): Comision de la URSS de los Asuntos de PNUMA. Colonizacion de los territórios áridos y lucha contra la desertification: enfoque integral. Centro de los Proyectos Internacionales—GKNT, Moscu, pp 17–24 (Capitulo II)

Chapter 6
Water Management in Drylands: Susceptibility and Risk of Desertification

Abstract The management of hydrographic basins and water resources must consider the environmental susceptibility and the risks of sanitization of soils around the world, especially in Drylands and irrigated perimeters. Soil salinization and sodicity are among the main problems of environmental degradation on the planet and can cause or increase desertification. Furthermore, water quality, lithogenetic salts contained in rocks, sediments, and soils, associated with inadequate irrigation management, can cause serious salinization problems of different degrees and intensity in Areas Susceptible to Desertification. Therefore, the transposition of hydrographic basins in drylands must follow conservationist precepts, with socioeconomic inclusion, focusing on water management at risk of desertification.

Keywords Watershed management · Water management · Environmental susceptibility · Risk of salinization

6.1 Transposition of Hydrographic Basins and Desertification

Drylands with critical evaporation rates favor a poor water balance. However, when groundwater exists, these are well sheltered, with a considerably smaller portion of water loss than that presented by surface springs. Groundwaters are the principal water reserves, especially during droughts, for priority human use, watering animals, and other uses. Therefore, in any scenario, the demands for this purpose must be guaranteed to conserve and evenly distribute this vital resource's contributions in minimum quantities and qualities.

The watersheds' transposition in the world promotes the connection of different drainage systems between climatic zones that may differ. Wetlands can provide significant water inputs to dry areas, especially semi-arid and dry sub-humid areas. This is already known worldwide from several transpositions between hydrographic basins of different sizes. It can be noticed, for example, in Africa (Pereira and do Nascimento 2016) and South America (Cadier 1996).

© The Author(s), under exclusive license to Springer Nature Switzerland AG 2023   61
F. Rodrigues do Nascimento, *Global Environmental Changes, Desertification and Sustainability*, SpringerBriefs in Latin American Studies,
https://doi.org/10.1007/978-3-031-32947-0_6

The production of fruit crops in some irrigated areas, for example, maintains a direct relationship with global markets. However, salinization and sodification induced by mismanagement of irrigation constitute a severe threat that involves agropoles. The high water wasting degree and soil degradation in irrigation compromises the generation of wealth and well-being in the countryside and can, as already seen, cause soil salinization (Castro and Santos 2020).

It should be, however, noted that much of the debate today about water supply in drylands focuses on the transposition of hydrographic basins. Many of these projects foresee sufficient average flow for rural supply (human consumption and diffuse agricultural uses), urban and industrial supply, and stimulating the development of irrigated agriculture with high added value.

It is also stipulated that the irrigated areas would be expanded, the agroindustry and the mining-metallurgical industry dynamized, and other modern services strengthened. Technically, these projects may involve a succession of channels, tunnels, reservoirs, and aqueducts that will originate from water intakes downstream of the source dams. The existence of branches, with sets of axes, and diversified sources of water abstraction, expands the wetlands.

However, it is notable that transpositions are selective and excluding besides the benefited economic sectors and portions of the territory—such as those of agribusiness. It favors big capital and some regions with high demographic densities, to the detriment of small producers and small and medium-sized cities in drylands (do Nascimento 2013).

Even if the proportion of average flow from the outflow to a diversion basin at the derivation points is modest, in addition to the inherent complexity of a diversion project in technical and political terms, other interests involved can be conflicting. Therefore, topics related to negotiation and conflict resolution for a climate of water justice and peace, economic-financial approaches, proposals for project preparation, institutional aspects, and physical-ecological and social issues are gaining momentum.

In this regard, there is a concern with drylands (arid, semi-arid, and dry subhumid). Furthermore, it is essential to know more about the impacts on desertification, the risks of salinization by irrigation, and a possible worsening of conflicts over land and water use. Depending on water quality, irrigation management, and evaporation rates, desertification by soil salinization can be a major local and regional problem.

Consequently, the most common problems affect the soil–water-plant relationship, which may exceed the carrying capacity of natural resources, according to Ayers and Westcot (1991). According to these authors, water-soluble salts are transported by irrigation and deposited in soils and plants, and they accumulate as evaporation occurs or are consumed by crops. Furthermore, watersheds, like other regions, may suffer effects resulting from the non-technification of the territory and the allocation of resources in areas not considered strategic for the business and industrial sectors, according to the perspective built by political decision-makers with their companies in the Areas Directly Affected by the Transposition.

6.2   Water Management and Risk of Desertification

Throughout human history, in addition to supplying food bases, irrigation management has caused problems such as chemical contamination of soil and water, which depletes or exceeds the carrying capacity of renewable natural resources, producing chemical, physical, and biological changes.

Zoon and Orlovski (1987) say that alkalization and secondary salinization—that is, that caused by irrigation—are recorded in history as representative anthropogenic factors of desertification. Proof of this is the soil conditions of present-day Iraq (Mesopotamia, 2400 a.C). Add to this the chemical contamination from agriculture and the intensification of production through fertilizers, insecticides, herbicides, and other chemical products, which can also contaminate animals and humans. This places agro-industrial complexes as potentially causing desertification due to extensive deforestation, erosion, and soil salinization.

The origin of salt is marine, lithic, or anthropogenic; however, soil salinity is one of the most worrying factors in modern agriculture, in which inadequate irrigation management is primarily responsible for compromising the quality of degraded soils and the abandonment and incorporation of land. Other problems are associated with drainage conditions, the piezometric level, and the concentration of salts in soils and groundwater (Richards 1954; Ayers and Westcot 1991).

According to Richards (1954), saline soils are more common in semi-arid and arid regions and are practically non-existent in humid regions. Nevertheless, not infrequently, in the drylands, saline soils typical of semi-arid and lowland regions are found, presenting high concentrations of salts, mainly sodium chloride. Salts are superficialized in the soil solution and concentrated after evaporation, which can be increased by irrigation (Mantovani 2003). Among the most common salts in irrigation water are sulfates, chlorides, carbonates, and bicarbonates, associated with the elements sodium, calcium, magnesium, and potassium.

Taking as an example the semi-arid climate that occurs in areas of high pressure, between the parallels of 20° and 30°—in both hemispheres, they are characterized by high temperatures, low annual thermal amplitude, and scarce and poorly distributed rainfall, with long periods of drought in the regional space. Due to the prevailing climatic conditions in these regions, the environmental susceptibility is considerable, composing areas with risks of droughts. These areas present values above the average temperature of 30 °C (which can reach extremes of 36 °C), and high thermo-pluviometric averages, above the evaporation of 1800 mm/year and potential evapotranspiration of 850 mm, with an aridity index above 46%, in addition to presenting differences in permoporosity of crystalline terrains and sediments, causing concentration of chemical substances. Added to this is the condensation of clouds formed over the oceans and composed of salts, which, when in contact with the terrestrial surface, give rise to edaphoclimatic conditions with soils with high levels of soluble salts and exchangeable sodium.

In principle, all waters and soils contain dissolved salts, being related to the mineralogical and chemical compositions of rocks, sediments, and soils, according to biogeochemical cycles (Drew 1986), with a potential for salinization in rainy and/ or drainage conditions, which can be aggravated by the use of saline and sodic waters for irrigation (Mesquita 2005).

However, soil salinization is the accumulation of soluble salts in the arable layers of the soil, preventing the development of most plants, caused by osmotic pressure, problems of intoxication, and loss of physical characteristics of the soil, including its fertility (do Hammecker et al. 2012; Salvati and Ferrara 2015). This type of salinization is natural/primary or human/secondary. Primary salinization is the accumulation of salts caused by natural characteristics, such as the high content of salts in the source materials, rise of soluble salts by capillarity, and low areas that receive materials from the surroundings by subsurface drainage (lateral and superficial). These elements may add to the high evapotranspiration rates of drylands in general (UEC 2009; Vasconcelos 2014; Salvati and Ferrara 2015; Castro et al. 2019).

Anthropogenic or secondary salinization is caused by human actions favoring the accumulation of soluble salts, such as using water with high salts, over-irrigation and/or absence of drainage systems, and excess fertilizers (UEC 2009; Salvati and Ferrara 2015).

Among the most saline soils in the drylands are quatzarenic neosols, planosols, and salic gleisols, but the salting of soils not originally halomorphic can occur in irrigated perimeters, increasing the environmental susceptibility to desertification. Soil salification may be a very complex reality that deserves special attention from technicians, civil society, and political decision-makers, without forgetting that without the exorheic runoff from the drainage of hydrographic basins in Drylands, which flow directly into the oceans, soil salting would be widespread.

With surface runoff or percolation, salts concentrate in surface and groundwater. As already said, salts dissolved in water are transported by irrigation and deposited in soils and plants, where they accumulate at the rate of evaporation or consumption by crops. Ayers and Westcot (1991) claim that the most common derived problems, which affect the soil–water-plant relationship, which may exceed the carrying capacity of natural resources, are:

- Salinity—Soil and water salts reduce water availability, affecting crop yields. The rise of the water table leads salts above the root zone, constituting additional sources of salts;
- Water Infiltration—Relatively high levels of sodium, or low levels of calcium in soil and water, reduce the rate at which irrigation water passes through the soil surface. This reduction can reach such magnitude to the point that the roots do not receive enough water with the irrigations;
- Specific ion toxicity (calcium—Ca^{++}; magnesium—Mg^{++}; sodium—Na^+; potassium—K^+; chloride—Cl^-; bicarbonate—HCO_3^-)—Certain ions—such as sodium, chloride, and boron (B)—contained in soil or water, accumulate in plants in high concentrations, which can cause damage and reduce yields of sensitive crops and plants, causing biological disorders such as necrosis, increased production of dry matter and late blight; and

- Other Problems—Excess nutrients reduce crop yields and/or their quality. Stains on fruit or foliage affect the marketing of products. Excessive corrosion of equipment increases maintenance and repair costs. Biological disturbances occur in plants and changes in water pH.

Pimentel et al. (2003) observed that irrigation increases food production, influencing labor occupation; however, business groups often accumulate profits. Furthermore, the impacts of the technologies used in the agropoles contained in the ASDs are still unknown.

Given this, efficient water use is one of the biggest agricultural challenges. Furthermore, inadequate agricultural practices and techniques used in planting can harm soil quality (structure and fertility), compromising productivity and requiring more resources linked to time, capital, and energy for food production. Besides, requiring, if necessary, prevention of desertification, without forgetting that, usually, field workers handle agrochemicals (fertilizers and pesticides) without proper safety measures, such as using protective equipment and technical guidance, with adverse effects on their health.

In general, maximum economic-financial efficiency can be followed by minimum efficiency of social and ecological well-being amid the optimal combination of inputs in the production process, producing entropies and competing wastes to exceed the geo-environmental support capacity. Although there are water meters in many agropoles, the techniques used are considered aggressors to the environment, mainly soil and water.

Doorenbos and Kassan (1994) pointed out that once the soil and agroclimatic conditions have changed, the water factor can influence the plants' water demands, the supply of this liquid, and each crop yield as the irrigation schedule and the quality of production. Undoubtedly, the main effect of salts on soils and plants is to alter the osmotic pressure capacity of plants, which hampers the absorption of water and nutrients from the soil (Richards 1954; Bohn et al. 1979; Moreira et al. 2014).

The water distribution in the soil profile is hampered by high levels of salts, especially sodium, producing temporary flooding. Soils can still become sterile in small or large areas, causing degradation of natural resources and desertification. Using good quality water containing low levels of salinity and risk of sodicity does not pose significant risks to irrigation (Richards 1954). However, with inadequate salt balance, as a result of drainage problems, irrigated areas can gradually increase the amount of salinized soils to different degrees (Cordeiro 2003 in Mesquita 2005).

In this scenario, the risk (in general) should be well assessed for this environmental issue. The risk would be the probability of occurrence of processes in time and space without frequency and determination. However, they affect, in some way, social life. Moreover, natural risk links to the probability of a natural disaster occurring. Therefore, for studies that involve the analysis of environmental risks, it is essential to consider the probabilities, crossing different variables in the estimation to determine which areas are more likely to materialize the risk (Brüseke 1997; Veyret and Meschinet de Richemond 2007).

6.3   Susceptibility and Risk of Salinization in a Watershed in the Drylands

Salinization is among the most severe environmental problems that cause land degradation, causing profound losses to soil agricultural production, including affecting the soil's capacity to store CO_2 (Brady and Weil 2013; do Nascimento 2020). These damages caused by land salinization reduce arable areas, impacting food production (Zewdu et al. 2016).

Soil salinization processes occur in several countries (Hammecker et al. 2012; Salvati and Ferrara 2015; Gkiougkis et al. 2015; Zewdu et al. 2016; Ma 2017), threatening the sustainability of agricultural productivity in arid and semi-arid regions (Korkanç and Korkanç 2016; Castro and Santos 2020) and food security.

Due to the conditions of hydroclimatic contingencies of the drylands, the environmental susceptibility is high, with equally dangerous environmental risks. Environmental susceptibility is the probability of a given environmental event occurring to the detriment of the combination of environmental characteristics and human interventions in a given area (Conoscenti et al. 2014; Patriche et al. 2016). Furthermore, the natural risk is linked to the probable occurrence of natural disasters (Brüseke 1997; Veyret and Meschinet de Richemond 2007).

The similarities of semi-arid regions always involve climatic, water, and phytogeographic aspects. There are low levels of humidity, scarcity of annual rainfall and rainfall irregularity over the years; long periods of water shortage; soils with physical and chemical problems, such as partially saline or carbonate soils; and absence of fluvial continuity, especially concerning autochthonous drainages. A fact common to semi-arid regions is an expressive dinoturnal thermal amplitude. Although the differences between minimums and maximums are modest, the results become expressive compared to average annual maximum and minimum values (Ab'saber 2003; do Nascimento 2015).

Concerning the salinization of soils and its implications on desertification, it is necessary to evaluate the relationship between risk × environmental susceptibility and this phenomenon and observe the environmental characteristics of each environment to be irrigated, together with the probability of occurrence of natural disaster, via salinization and sodicity of the soils.

In South America and the Caribbean, there are three semi-arid cores: the dry arrheic diagonal of the Southern Cone, which crosses the Andean chain, extending toward 5° south latitude along with Argentina, Chile, and Ecuador; Guajira region in Venezuela and Colombia, with semi-arid regions on the Caribbean front, in the extreme northwest of the South American continental block. Among these semi-arid regions, the Brazilian semi-arid stands out—to constitute itself as one of the three cores of semi-arid regions in South America, in the context of a predominantly humid continental area.

The semi-arid region of Brazil covers almost one-eighth (12.5%) of the country's territory and occupies about 1 million km² (Conti and Furlan 1995). The dry Northeast of that country represents the most populated semi-arid region in the world, with about 27 million people, and perhaps, is the region with the most rigid agrarian structure on the planet (da Cunha Rebouças 2002; Ab'saber 2003). The region has the highest human fertility rate in the Americas. It is a region that generates and redistributes people, even under the pressures of droughts, poverty, and misery (Ab'saber 2003).

Mesquita (2005) studied the salinization of soils in the semi-arid region of Brazil. The author claimed that the hydrochemical water quality in the basins is directly related to soils' salinization and sodification. In general, the contribution of the upper course of the basin to increase salinity, compared to its lower course, can be accentuated or discreet. As an exemplification of the question, we highlight an intermittent seasonal basin in this semi-arid region of South America/Brazil (Fig. 6.1). The results showed no salinity increase, even in the months of greater aridity, in the analyzed samples, at least between the middle and upper reaches of the basin, considering the sampling points' distribution (Fig. 6.2).

The waters of the lower part of the test basin presented a higher salinity due to the higher chloride concentrations (78.9%) in the waters coming from the semi-arid region (crystalline), and 21.1% are bicarbonate—from and of springs with humid and dry sub-humid climate in sectors of Plateau with 800 mt of altitude. Chloride levels, however, are well below the toxicity limits for crops in general, so there are no restrictions on their use for irrigation.

In the upper reaches of the basin, the waters are mixed (magnesian, calcium, and sodium—$Na^+ + K^+$) in terms of cations (Na^+, Ca^{++}, Mg^{++}, K^+) and bicarbonated (HCO_3^-), sulfide (SO_4^-) and chlorinated (Cl^-), in terms of anions. By using the

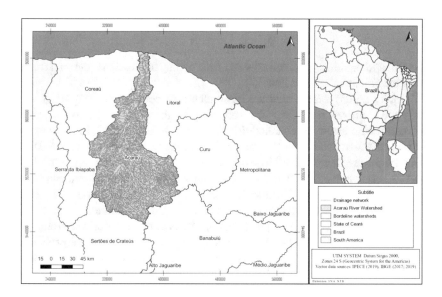

Fig. 6.1 Risks of salinization and desertification—sampling points distribution

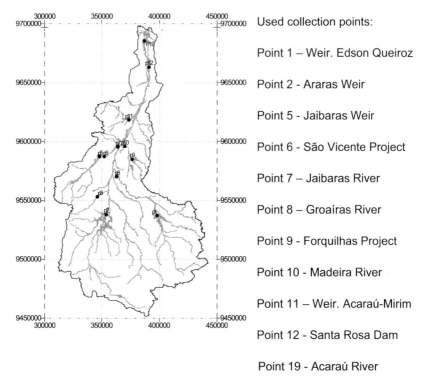

Used collection points:

Point 1 – Weir. Edson Queiroz

Point 2 - Araras Weir

Point 5 - Jaibaras Weir

Point 6 - São Vicente Project

Point 7 – Jaibaras River

Point 8 – Groaíras River

Point 9 - Forquilhas Project

Point 10 - Madeira River

Point 11 – Weir. Acaraú-Mirim

Point 12 - Santa Rosa Dam

Point 19 - Acaraú River

Fig. 6.2 Risks of salinization and desertification—sampling points distribution. *Source* Mesquita (2005)

Piper triangle, sodium predominates (57.6%) among the cations and bicarbonate among the anions (Mesquita 2005).

Mesquita (2005) interpolated the mean electrical conductivity (Fig. 6.3), identifying the distribution of the water suitability of the analyzed basin for salinity risks, comparing electrical conductivity (average values, at 25 °C, in decisiemens/meter, dS m^{-1}) and sodium adsorption ratio, in four classes: unrestricted—0.00 to 0.20; without restriction—0.20 to 0.40; without restriction—0.40 to 0.70; mild to moderate restriction—0.70 to 0.80. The author points out that the area characterization regarding its use for irrigation is uniform, without salinization risks, except for point 10 (Madeira River), which presents a slight to moderate restriction. This author also divided the categories that show the risks of problems arising from water sodicity (Fig. 6.4) into "no problems, increasing problems, and severe problems". The author verified that most of the basin has increasing problems regarding the risk of infiltration by the sodicity of the water. Points 5–7 have severe sodicity problems in water. Point 10 has growing problems, and the others displayed no problems.

Both risks of salinity and sodicity of water are directly associated with the soil in the agricultural system since water resources are managed by irrigation. Considering

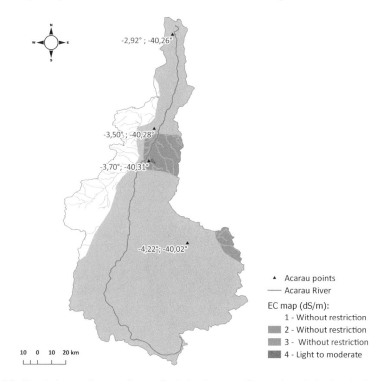

Fig. 6.3 Restriction on the use of water for irrigation according to the salinity in a hydrographic basin. *Source* Adapted from Mesquita (2005)

the integrated analysis of the environment, these risks of environmental chemical contamination are associated with the support capacity of geoenvironments in their potential and limitations.

The results prove that detailed investigations on the hydrochemical degradation of soils deserve attention in future works and interdisciplinary projects on desertification—a fact that can extrapolate to different basins in drylands worldwide. The salinization of soils induced by inadequate management of water resources is among the principal risks of desertification worldwide.

In another case, according to Castro et al. (2019), the combination of biophysical elements (slope, geology, and potential evapotranspiration, natural characteristics of each type of soil) with land use and intensive water use in agricultural production contributes to evaluating susceptibilities to soil salinization for mapping the different susceptibility levels to soil salinization. They studied areas in the Caatinga Biome, a semi-arid region of Brazil (Fig. 6.5). The authors concluded that more than 97% of the lands in the surveyed area were classified as having medium (59.71%), high (15.72%), and very high (21.84%) susceptibility to salinization. On the other hand, only 2.72% of the total area was classified as having low salinization susceptibility (Fig. 6.6).

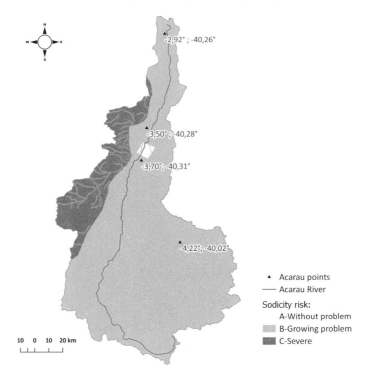

Fig. 6.4 Risk of infiltration problem along the basin promoted by water sodicity. *Source* Adapted from Mesquita (2005)

The decisive factors for these results were the presence of soils with a high propensity to salinization (Nátric Planosol, Haplic Gleissolo, and Fluvic Neosol), which are widespread in the Brazilian semi-arid region. The type of management applied in the production areas is characterized by using large amounts of water through furrow irrigation.

It is worth noting that it is necessary to improve hydrological indicators of degradation/desertification processes when several intensive studies on hydrological aspects in arid zones and monitoring of water resources are carried out on salinization by irrigation. Some hydrological indicators are recognized as effective in studying surface and groundwater, with an effective contribution in other case studies, as well as in other regions that protect arid, semi-arid, and dry sub-humid lands.

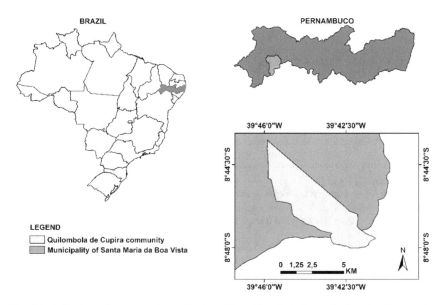

Fig. 6.5 Soil salinization assessed area. Caatinga Biome/Brazil, Quilombola Curupira Community. *Source* Castro et al. (2019)

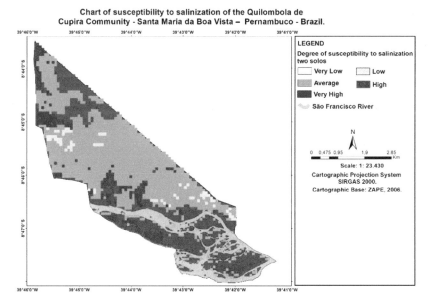

Fig. 6.6 Classification and degrees of susceptibility to salinization. *Source* Castro et al. (2019)

References

Ab'saber AN (2003) Os domínios de natureza no Brasil: potencialidades paisagísticas. Ateliê Editorial, São Paulo

Ayers RS, Westcot DW (1991) A qualidade da água na agricultura. Estudos FAO—Irrigação e drenagem. UFPB, Campina Grande

Bohn H, Mcneal BL, O'Connor GA (1979) Soil chemistry. Wiley, Brasil, USA

Brady NC, Weil RR (2013) Acidez, Alcalinidade, Aridez e Salinidade do Solo. In: Brady NC, Weil RR (eds) A Natureza e propriedades dos solos, Terceira ed., vol 518. Bookman, Rio de Janeiro, pp 76–97

Brüseke FJ (1997) Risco social, risco ambiental, risco individual. Ambiente and Sociedade, Campinas, [S.l] 1(1):117–134

Cadier E (1996) Hydrologie des petits bassins du Nordeste Brésilien semi-aride: typologie des bassins et transposition écoulements annuels small watershed hydrology in semi-arid north-eastern Brazil: basin typology and transposition of annual runoff data. J Hydrol 182(1–4):117–141

Castro FC, Santos AM (2020) Salinity of the soil and the risk of desertification in the Semiarid Region. Mercator (Fortaleza, Online), vol 19, pp 1–13

Castro FC, Araujo JF, Santos AM (2019) Susceptibility to soil salinization in the quilombola community of Cupira—Santa Maria da Boa Vista—Pernambuco—Brazil. CATENA 179:175–183

Conoscenti C, Angileri S, Cappadonia C, Rotigliano E, Agnesi V, Märker M (2014) Gully erosion susceptibility assessment by means of GIS-based logistic regression: a case of Sicily (Italy). Geomorphology 204:399–411

Conti JB, Furlan SA (1995) Geoecologia: o clima, os solos e a biota. In: Ross JLS (Org) Geografia do Brasil. Edusp, São Paulo, pp 67–208

da Cunha Rebouças A (2002) Água doce no mundo e no Brasil. In: da Cunha Rebouças A, Braga B, Tundisi JG (Org). Águas doces no Brasil: capital ecológico, uso e conservação, 2nd edn. Escrituras, São Paulo, pp 1–37

do Nascimento FR (2012) Os recursos hídricos e o trópico semiárido no Brasil. Geographia 14:82–109

do Nascimento FR (2013) O Fenômeno da Desertificação. Cegraf, Goiânia

do Nascimento FR (2015) Os semiáridos e a desertificação no Brasil. Rev Eletrôn Prodema (REDE) 9:7–26

do Nascimento FR (2020) Manejo de água em bacias hidrográficas e desertificação—gestão e ações planejadas para zonas tropicais. In: Figueró e Cláudio de Mauro A (Org). Governança da água: das políticas públicas à gestão de conflitos, vol 1, 1st edn. EPTEC, Campina Grande, pp 232–246

do Nascimento FR et al (2005) Geo-environmental analysis and identification of degraded áreas susceptible to desertification in a semiarid, tropical ecozone: the Acaraú river basin in Northeastern Brazil. Rev Soc Natureza Esp 361–368

Doorenbos J, Kassan AH (1994) Efeito da água no rendimento das culturas. In: Estudos FAO—Irrigação e Drenagem, vol 33. UFPB, Campina Grande

Drew D (1986) Processos interativos homem-meio ambiente (Trad. dos Santos JA, Bastos S). Difel, São Paulo

Gkiougkis I, Kallioras A, Pliakas F, Pechtelidis A, Diamantis V, Diamantis I, Ziogas A, Dafnis I (2015) Assessment of soil salinization at the eastern Nestos River Delta, N.E. Greece. CATENA 128:238–251

Hammecker C, Maeght JL, Grunberger O, Siltacho S, Srisruk K, Noble A (2012) Quantification and modelling of water flow in rain-fed paddy fields in NE Thailand: evidence of soil salinization under submerged conditions by artesian groundwater. J Hydrol 456/457:68–78

Korkanç SY, Korkanç M (2016) Physical and chemical degradation of grassland soils in semi-arid regions: a case from Central Anatolia, Turkey. J Afr Earth Sci 124(579):1–11

Ma L, Ma F, Li J, Gu Q, Yang S, Wu D, Feng J, Ding J (2017) Characterizing and modeling regional-scale variations in soil salinity in the arid oasis of Tarim Basin, China. Geoderma 305:1–11

Mantovani W (2003) Degradação dos biomas brasileiros. In: Ribeiro WC (Org) Patrimônio ambiental do Brasil. Edusp: Imprensa Oficial do Estado de São Paulo, São Paulo, pp 367–442

Mesquita TB (2005) Caracterização da qualidade das águas empregadas nos distritos irrigados da bacia do Acaraú. 2004. 62 f. Dissertação (Mestrado em Irrigação e Drenagem)—Universidade Federal do Ceará, Fortaleza

Moreira LCJ, Teixeira AST, Galvão LS (2014) Laboratory salinization of Brazilian alluvial soils and the spectral effects of gypsum. Remote Sens 6:2647–2663

Patriche CV, Pirnau R, Grozavu A, Rosca B (2016) A comparative analysis of binary logistic regression and analytical hierarchy process for landslide susceptibility assessment in the Dobrovat River Basin, Romania. Soil Sci Soc China 26:335–350

Pereira JI, do Nascimento FR (2016) Susceptibility to desertification in Chicualacuala, Republic of Mozambique. Int J Geosci 07:229–237

Pimentel CRM, de Souza Neto, de Freitas Rosa M (2003) Aspectos econômicos dos perímetros irrigados: Curu-Paraipaba, Curu-Recuperação, Araras Norte e Baixo Acaraú. Relatório de pesquisa. Embrapa/CNPAT, Fortaleza

Richards LA (ed) (1954) Diagnosis and improvement of saline and alkaline soils, vol 60. USDA—Agricultural Handbook. Washington, DC, USA

Salvati L, Ferrara C (2015) The local-scale impact of soil salinization on the socioeconomic context: an exploratory analysis in Italy. CATENA 127:312–322

UEC (2009) European Commission's Science. Salinização e sodificação. In: Ficha 646 informativa, vol 4. UEC, Madrid

Vasconcelos MCCA (2014) Salinização do solo em áreas irrigadas: Aspectos físicos e químicos. Agropec Cient Semiárido (ACSA) 10(1):20–25

Veyret Y, Meschinet de Richemond N (2007) O risco, os riscos. In: Veyret Y (Org) Os riscos: o homem como agressor e vítima do meio ambiente. Contexto, São Paulo, pp 23–79

Zewdu S, Suryabhagavan KV, Balakrishnan M (2016) Land-use/land-cover dynamics in Sego Irrigation Farm, southern Ethiopia: a comparison of temporal soil salinization using geospatial tools. J Saudi Soc Agric Sci 15:91–97

Zonn IS, Orlovski NS (1987) Factores antropogênicos de la desertificación. In: Programa de las Naciones Unidas para el Médio Ambiente (PNUMA): Comision de la URSS de los Asuntos de PNUMA. Colonizacion de los territórios áridos y lucha contra la desertificación: enfoque integral. Centro de los Proyectos Internacionales—GKNT, Moscu, pp 17–24 (Capitulo II)

Chapter 7
Agenda in Addressing Global Environmental Changes and Desertification

Abstract Global environmental changes have taken place without scales and precedent since the beginning of the Holocene, threatening the biosphere and humanity and sometimes exceeding the planetary boundaries. The climatic extremes for drought, high temperatures, floods, torrential rains, melting glaciers and even heavy snowfalls, and changes in land use, air, water, and soil contamination with the Incorporation of new entities, have increased their frequency and intensity. Phenomena that were once rare-disturbing now occur on a scale of a tenth of years or even more often. Human actions of land use, deforestation, and greenhouse gas emissions are among the main factors that produce global environmental changes, triggering a crisis and accelerating climate change. Desertification is one of the main consequences of these changes on global scales, which affect climates and, in particular, the poorest countries and regions of the planet. The global environmental order, with the unrestricted support of the UN, puts the environmental agenda on red alert, with an emphasis on climate, as a challenge for nations.

Keywords Global environmental challenges · UN · Environmental and climate change · Desertification

7.1 Extent of the Problem and Challenges for an Environmental Agenda

Humanity and its international relations have one of the most significant challenges in history with new scenarios for the twenty-first century in the face of global environmental changes, especially the climate crisis. Among these problems, desertification stands out. The increase in the planet's temperature by 1.1 °C compared to pre-industrial levels already causes profound global impacts, with regional repercussions, causing extreme events such as waves of calluses, droughts, floods and heavy snowfall, contamination of seas and oceans, mass international migrations by climate and environmental refugees. At the same time, arable land is degraded, followed by an accelerated pace of land incorporation by increasing deforestation.

F. Rodrigues do Nascimento, *Global Environmental Changes, Desertification and Sustainability*, SpringerBriefs in Latin American Studies, https://doi.org/10.1007/978-3-031-32947-0_7

Water and food security reach their thresholds around the world. Desertification is one of the possible causes of the fall in land productivity. According to the Global Biodiversity Outlook, 45% of the food produced worldwide comes from dry regions. Drops in productivity, and lack of food and water, are generating insecurity worldwide. Two out of five people already feel the phenomenon's effects (ONU 2021). Pandemics have been recurrent since the last century, usually linked to ecological and sanitary conditions. A world of challenges and new challenges around the world require adaptations to emerging scenarios of environmental changes at ever faster scales.

> The citizens of the (…) world are experiencing an epidemic without scale, magnitude, and precedent: coronavirus, COVID-19, SARS-CoV-2. We are living at the "corner of history," as we are experiencing another moment, other experiences in today's world and its civilizing process. A series of other circumstances never occurred until now, in terms of simultaneity of information in time and space. (do Nascimento 2020, p. 141)

Therefore, the international environmental order focuses on governance of the international environmental order without scales or precedents.

7.1.1 Global Actions and International Institutions: Practices with Local Repercussions

Proactive policies and investments of financial resources with emphasis on scales of regional and local actions by continent and country would be essential to distinguish areas by technical studies identifying nuclei of desertification, where areas heavily affected by climate change exist.

We would also say that the mitigation actions and policies discussed here align with the Millennium Declaration, Agenda 21. The Millennium Development Goals— MDGs guide them and their relationship with the Sustainable Development Goals (ODS, of the UN). The SDGs have 17 Goals and 169 targets, created to eradicate poverty and promote a dignified life for all within the conditions our planet offers and without compromising the quality of life of the next generations.

The SDGs of the 2030 Agenda (2016 to 2030—Transforming Our World: The 2030 Agenda for Sustainable Development, A/70/L.1) rest on what the MDGs established to respond to new challenges. It is an action plan for people, the planet, and prosperity. All 17 goals are directly or indirectly related to global changes and desertification. Goal 13: "Take urgent action to combat climate change and its impacts" stands out. Indicators monitor the goals; the results of each nation can be compared, offering a global panorama for the follow-up of the Agenda by the UN around the world.

The SDGs are essential in the search for scientific and technological solutions to support decision-makers in the collective construction of local agendas to improve the inhabitants' quality of life. Although global changes affect the planet, priority is given to treating desertification on a regional and local scale to guarantee access to vital goods and services such as drinking water and the maintenance of local socioeconomic activities based on rural production.

In addition, the SDGs provide in their Art. 15.3 actions to recover degraded areas, combat desertification, and measures to achieve neutral land degradation. In its Nationally Determined (NDC) contribution to the Climate Change Convention, each country has an ethical-ecological obligation to create goals for the recovery of forests and degraded pastures and to implement Innovation integrating Crop, Livestock, and Forestry/ILPF, which a percentage should be implemented in Biomes represented by drylands (Bungenstab et al. 2019). In this regard, many biomes in the Americas, Africa, Asia, and Oceania have significant land stocks available for conversion (Soares-Filho et al. 2014; Bouchle et al. 2015).

It is important to remember that the 26th conference of the parties to the UN Framework Convention on Climate Change 2021 (COP26—1 and 11/12/2021, Glasgow, Scotland) also added the 15th meeting of the parties to the Kyoto (CMP16) and the second meeting of the parties to the Paris Agreement (CMA3). The UN (2022) final document presents advances, technical, political, and ethical hopes, besides uncertainties about the GHG emissions reduction: polluting countries such as India and China ask for an adjustment of the text, placing "reduction" instead of "abandonment" of the coal use. In addition to these contradictions, there are achievements in reducing deforestation and emissions of these gases.

Thus, developed countries ratified the promise of the Paris Agreement to stop US$ 100 billion/year so that developing nations can face the effects of climate change. On the other hand, poor and developing countries have already asked for losses and damages to be paid by rich nations to offset losses from climate change. In this universe, it is necessary to promote a large fund and global plan with an Action Program to Combat Desertification and Mitigation of the Effects of Drought.

At the same time, there are complex and diverse difficulties that persist. A survey by the World Health Organization/WHO shows that only 25% of countries were able to put into practice strategies to protect the population from the effects of this problem. In another direction, 77% of nations have national plans in this regard but do not have the funds.

About desertification, there are several connections within COP 26: Among the poorest countries, there are those in drylands. Moreover, with a tendency to increase extreme droughts and advance of areas in the process of desertification, climate change is associated with this phenomenon in an imbricated way. Therefore, all anticipatory actions to mitigate climate change represent, in practice, the fight against desertification.

It is crucial to remember that poor and developing countries have already claimed "losses and damages" to be paid by rich nations to compensate for losses from climate change. Therefore, financial investments in regional and local actions would be fundamental in identifying Desertification Nuclei, where Areas Heavily Affected by Climate Change are the focal points.

The poorest countries are located in the drylands. Furthermore, extreme droughts and extension of areas in the process of desertification due to land use and climate change tend to increase in these areas (FAO/ITPS 2015; ONU 2021).

Concerning policy recommendations with practical consequences, we suggest, without fail, monitoring and applying the complete list of indications in the holistic treatment of natural resource management, with emphasis on desertification, given global changes. Since the Stockholm/Sweden Conference, 1972, passing all the most essential environmental conventions and treaties in the world since then, for example, RIO'92/Brazil, 1992; Kyoto/Japan Protocol, 1997; Rio+10/South Africa, 2002; Rio+20/Brazil, 2012; Paris/France Agreement, 2016; COP26/Scotland, 2021. These events bring within them valuable documents, from time to time, with significant actions suitable for dealing with the issue, and even if there is a large gap between theory and practice, it is necessary to insist on these actions: poverty reduction and social inequality; sustainable expansion of production capacity; preservation, conservation and management of natural resources; democratic management; and institutional strengthening.

Regarding desertification, it is imperative to follow the *Treaty on arid and semi-arid zones*, one of the 46 documents prepared by NGOs from five continents gathered in the parallel forum to Eco-92. This treaty states that Drylands are complex ecosystems with sufficient natural potential to provide a good quality of life for their populations, provided that a concept of socially fair, ecologically sustainable, and culturally appropriate development is adopted (do Nascimento 2013).

These initiatives should involve a range of policy decision-makers, including national governments, international organizations, the private sector, and civil society, working together to stop land degradation. Again, international partners stand out, such as the United Nations Convention to Combat Desertification (UNCCD), the Food and Agriculture Organization of the United Nations (FAO), the World Bank (WB), the Global Environment Facility (GEF), the European Union, and the International Union for Conservation of Nature (IUCN), which mobilize substantial investments and implement initiatives in different adaptation scenarios (UN/UNCCD 2020; UN 2021).

7.2 New Environmental Scenarios and Common Challenges

Planetary Boundaries/PB deal with global sustainability and define boundaries for humanity so that life can happen safely. Exceeding any of these Limits could be catastrophic, as it would cause non-linear and abrupt environmental changes in terrestrial systems. The PBs are climate change; Biosphere Integrity; Changes in land use; Biogeochemical Flows; Depletion of Atmospheric Ozone; Use of fresh water; Ocean Acidification; Loading of atmospheric aerosols; e., Incorporation of New Entities (Rockström et al. 2009a, b; Steffen et al. 2015; SRC 2022).

In this way, new global environmental challenges, given environmental and climate changes, with emphasis on desertification, are a priority agenda of the new World Environmental Order. Nevertheless, unfortunately, the pace of production and consumption of global society during the COVID-19 pandemic influenced little the planet's environmental directions in proactive actions beyond the agreements made at the end of 2021 at COP26.

With the course of the pandemic and its future amortization, economic growth must be guided by environmental sustainability, as never before. The non-compliance with this approach may expose to the risk of new pandemics, increased poverty, social exclusion in extreme weather scenarios, and increased water, energy, and food insecurity, which tend to be increasingly harmful.

It is also essential to consider that there is less additional agricultural land for use than is assumed, considering restrictions and trade-offs (a range of restrictions and trade-offs associated with land conversion). Second, land conversion is always associated with high social and ecological costs. For this reason, it must be considered that the expansion of agricultural production will face a complex scenario of competing demands and compensations. It is also considered that only potentially available crops implying low ecological and social costs with conversion to crops should be included (Lambin et al. 2013).

It can be said, therefore, that there are central themes and common challenges, even with Planetary Boundaries. Despite these, the SRC (2022) claims to know about the problems and possible solutions. Quick and bold actions are needed by Governments, focusing on alternative energy.

Several typologies can be proposed to classify environmental scenarios, respecting their nuances and differences (Pulver and van Deveer 2007) for each continent in the face of global and climate changes. Alcamo (2008) need to be addressed, reviewing the political and governance crises on Climate Change (Park et al. 2008; UNESCO 2021).

- Promote sustainability in the management of natural resources, notably climatic, biological, water, and soil resources, at multiple scales (Biggs et al. 2007).
- Highlight the importance of soil resources worldwide, their management, and their role in ecosystem processes. According to Lannetta and Colonna (2022) by avoiding salinization with responsible governance and policies (FAO/ITPS 2015).
- Assess functional scenarios of global change in their current problems and challenges (Parson 2008).
- Assess vulnerability to global environmental changes (Adger 2006).
- Monitor new scenarios and their impacts and changes in sustainability (UNEP 2012). Also stimulating the perception of climate change (Hansen 2012).
- Assess climate change, focusing on adaptations to this reality (Obermaier and Rosa 2013) in the face of uncertainties.
- Mitigate ecosystems and human well-being (Raskin et al. 2005), also considering: Integration of local and scientific knowledge for environmental management (Raymond et al. 2010).
- Link future scenarios, between society and ecology, at scales by multi-scale scenarios (Biggs et al. 2007).

- Promote bottom-up political and environmental decision-making, where agricultural management versus environmental protection is the keynote (Lambin et al. 2013).

Another critical action level is to seek inter-institutional partnerships for scientific improvement and governance (UNESCO 2021). At the global level, seeking greater cooperation between countries is essential. In this effort, the German Agency for Technical Cooperation (*Deutsche Gesellschaft fuer Technische Zusammenarbeit, GTZ*) stands out, with the funding of around tenth projects in several affected countries, as well as the German Service for Technical and Social Cooperation (DED) and the Konrad Adenauer Foundation, with the improvement of studies and the fight against degradation, which can culminate in desertification.

In this way, we cite the major works of the *Centro Del Agua para Zonas Aridas y Semiáridas de América Latina y el Caribe* (CAZALAC), *Intergovernmental Hydrological Program for Latin America and the Caribbean* (PHI-LAC) (UN/UNESCO), Red G-WADI-LAC (*Red Global para el Agua y Desarrollo de la Información en las Zonas Aridas, in its various segments,* Institute Deltares, AGWA/Alliance for Global Water Adaptation.

Other global networks of institutions should be remembered, such as higher education institutions promoting academic exchange and cooperation in the area of natural resource management, especially related to water, ecosystem, land, renewable energy resources, and climate change. Among these, the Centers for Natural Resources and Development (CNRD) stands out (https://www.cnrd.info/ 2022).

All this is crucial to work with professionals from different parts of the world and multidisciplinary formations who are experts in the environment, natural resources, climate, and environmental changes. In this way, local actions and sustainability gain strength by mitigating global impacts.

7.3 Global Impacts, Local Actions and Sustainability

In the context of global environmental changes and advancing desertification, climatic conditions and their hydrological reflexes offer risk and increasing demand for risk management for water security in watersheds on several continents, causing degradation and significantly affecting the sustainability of rural communities in the face of climate changes and variability (droughts and floods). In addition, severe systemic and cyclical problems of droughts and/or floods affect production and populations, providing minimal water inputs. This matter requires the practice of articulated regional-local strategies, continuous planning of droughts (management, monitoring, mitigation of drought stages), and water and soil management, with the recovery of degraded areas.

In the rural areas, it is crucial to encourage small producers, strengthen local and regional infrastructure, and spread agroecology. Agroecology is a productive model of family farming that must be associated with water resources and has a lower demand for external inputs to the property. Besides, organic agriculture increases production and protects natural resources, presenting fewer risks in case of droughts and market changes. It is cheap, profitable, and commercial agriculture. It can, among other aspects, use plants in symbiosis with biological nitrogen fixation bacteria (BNF) and endomycorrhizal fungi.

It is crucial to expand credit policies for family farming, providing resources for small producers to invest in the modernization of their agroextractive, forestry, and fishery activities to achieve this goal, besides encouraging associative and cooperative behaviors, followed by a profound agrarian reform, as medium- and long-term measures. In the short term, the producer must invest and pay in quotas from Seasonal Crop Plans in each country and region.

Technically, it is urgent to value the potential for conservation of vegetation and/ or its (re)forestation to avoid excessive evapotranspiration. Associated with this, the development of hydrological studies can help even more in the treatment of land degradation and in living in the semi-arid region. Therefore, recommending measures to combat desertification, mainly with four focus of studies, as directed by Batchelor and Wallace (1995), should be:

- to verify anthropogenetic effects on floodplain occupation and overgrazing, based on *feedback* from hydrological and surface-atmosphere factors;
- improve some degradation indicators, mainly with remote sensing studies, to certify the extent of the desertification phenomenon;
- investigate the water use increase and its efficient consumption by vegetation and crops. These studies are essential to understand better the water-soil–plant relationship and agricultural development in environments that suffer from aridity, requiring balancing the minimum amount of water needed for vegetables and the optimization of agriculture; and
- finally, to associate the management of natural resources with hydrological, agronomic, and socioeconomic potential studies, based on a holistic approach to protect and, possibly, reverse land degradation in the semi-arid region.

It is expected that the space organizers put into practice the precepts of Agenda 21, especially Chap. 18, which deals with the protection, quality, and water resources supply, linked to Chap. 12.2, which considers the semi-arid region and desertification.

These chapters advocate the fact that it is necessary to integrate measures for the protection and conservation of water sources; develop techniques for public participation in decisions; mobilize water resources, especially in arid and semi-arid areas; and develop water supply (desalination, reuse, and replacement of aquifers, etc.). Furthermore, it is also crucial that, due to the fight against desertification and drought, the degradation of biological exploitation associated with socioeconomic activities on water resources must be considered in managing fragile ecosystems. In particular, the superficial hydrological degradation is due to the loss of vegetation cover, and the degradation of groundwater is due to changes in recharge conditions.

Besides, it is urgent to carry out agroclimatic zoning to assess agricultural aptitudes and to properly plan the priorities of oriented crops of cultures according to their sub-regional and local climatic diversifications.

In this way, local development would dynamize the regional and local economies according to the optimal use of natural resources, improving the well-being of populations and social groups in the condition of commitment that elects space as a place of active solidarity, changing people's behavior and attitude.

It is necessary to review irrigation policies and techniques in the world's drylands. As a matter of urgency, given the high environmental susceptibility in irrigated agriculture, there is a need to monitor and control the salts used in irrigation to reduce the risks of degradation of renewable natural resources, with appropriate management for each landscape and associated ecosystem. Furthermore, the irrigation method and soil characteristics must integrate to avoid the risks of desertification by salinization, minimizing the compromise of agricultural production and crop profitability (do Nascimento 2013; ONU/ITPS 2015).

Understanding desertification as a complex and interactive environmental problem, however, establishing pari passu guidelines for each landscape unit is fundamental to maintaining environmental balance.

The sustainability concept must be expanded to propose a new treatment between nature and culture, founding a new economy, reorienting the potential of science and technology, and building a new political and environmental culture formulated based on sustainability ethics. Values, beliefs, and feelings must support all this local knowledge, which renews existential meanings, ways of life, and ways of inhabiting the planet so that proactive actions and interdisciplinary tasks are developed to treat desertification.

We know that achieving the SDGs, especially Goal 13 (urgent measures to combat climate change and its impacts), must seek to go beyond the simple increase in consumption and production of material goods and services. It is a political, institutional, and intellectual challenge that does not prove to be practically insurmountable, as has been followed by Sustainable Development itself since 1987. Even so, it is necessary to guarantee it through a process that allows individuals, communities, and governments to the redemption of rights and the autonomy to decide their future along with the common good. Even though this desire appears as "a midsummer night's dream", reality demands the realization of something more in this sense.

It is also important to emphasize that, in addition to macroeconomic ideas, the "*locus*"—that is, the area of community participation of populations and social groups in the wake of global environmental changes—must be considered in the development process.

Desertification must be considered a complex environmental problem, compromising the support capacity of the ecosystems that comprise geoenvironments. Such an approach should contribute to a new understanding of the world in facing humanity's challenges, even because a desertification approach is embedded in sustainability principles in the face of global environmental changes (such as the climate crisis), extrapolating the technical and polarized idea of the term. What is more, this problem must be considered from a multiple and diversified perspective, highlighting its importance for living with the phenomenon of drought and the consequent improvement of the population's quality of life. This perspective must invert the maxim from "Think globally and act locally" to "Think locally and act globally".

References

Adger N (2006) Vulnerability. Glob Environ Chang 16:268–281

Alcamo J (2008) Environmental futures: the practice of environmental scenario analysis, 1st edn. Elsevier, Amsterdam, 212p

Batchelor CH, Wallace JS (1995) Hydrological process, dryland degradation and integrated catchment resource management. In: Desertification control bulletin: a bulletin: of world events in the control of desertification, restoration of degraded lands an reforestation, vol 27. United Nations Environment Programme (UNEP), pp 27–34

Biggs R, Raudsepp-Hearne C, Atkinson-Palombo C, Bohensky E, Boyd E, Cundill G et al (2007) Linking futures across scales: a dialog on multiscale scenarios. Ecol Soc 12(1), art. 17

Bouchle R, Grecchi RC, Shimabukuro YE, Seliger R, Eva HD, Sano E, Archard F (2015) Land cover changes in the Brazilian Cerrado and Caatinga biomes from 1990 to 2010 based on a systematic remote sensing approach. Appl Geogr 58:116–127

Bungenstab DJ et al (2019) ILPF: inovação com integração de lavoura, pecuária e floresta. Embrapa, Brasília, DF, 835p

Centers for Natural Resources and Development (CNRD) (2022) Networking and sustainability. Cited on: https://www.cnrd.info/. Accessed: 15 Feb 2022

do Nascimento FR (2013) O Fenômeno da Desertificação. Goiânia, Cegraf

do Nascimento FR (2020) Uma análise na escala do Nordeste e estratégias Regionais na organização/integração para o combate ao SARS-COV-2. In: Ribeiro WC (Org) COVID 19. Passado, Presente, Futuro, vol 1, 1st edn. FFLCH/USP, São Paulo, pp 141–167

Food and Agriculture Organization of the United Nations-FAO/Intergovernmental Technical Panel on Soils (ITPS) (2015) Status of the World's soil resources. Main report, Rome, 650p

Hansen J, Sato M, Ruedy R (2012) Perception of climate change. Proc Natl Acad Sci USA 109:E2415–E2423

Lambin EF et al (2013) Estimating the world's potentially available cropland using a bottom-up approach. Glob Environ Chang 23(5):892–901

Lannetta M, Colonna N (2022) Salinização. In: Lucinda. Land care in desertification affected areas. Fascículo B, vol. 3. Cited on: http://www.icnf.pt/portal/naturaclas/ei/unccdPT/ond/lucinda/b3_booklet_final_pt_rev3. Accessed: 17 Feb 2022

Obermaier M, Rosa L (2013) Mudança climática e adaptação no Brasil: uma análise crítica. Estudos Avançados 27(78):153–176

ONU (2021) Nações Unidas: Degradação de terras impacta 3.2 milhões de pessoas no mundo. Cited on https://news.un.org/pt/story/2018/06/1627442. Accessed: 13 Dec 2021

Park J, Conca K, Finger M (eds) (2008) The crisis of global environmental governance. Routledge. Sterman JD (ed) Policy forum: risk communication on climate change: mental models and mass balance. Science 322

Parson EA (2008) Useful global-change scenarios: current issues and challenges. Environ Res Lett 3

Pulver S, van Deveer S (2007) Global environmental futures—interrogating the practice and politics of scenarios. Brown University

Raskin P et al (2005) Global scenarios in historical perspectives. In: Carpenter PLPSR, Bennett EM, Zurek MB (eds) Ecosystems and human well-being. Scenarios: findings of the scenarios working group of the Millennium ecosystem assessment, vol 2. Island Press, Washington, DC, USA, pp 35–44

Raymond CM, Fazey I, Reed MS, Stringer LC, Robinson GM, Evely AC (2010) Integrating local and scientific knowledge for environmental management. J Environ Manage 91(8):1766–1777

Rockström J, Steffen W, Noone K et al (2009a) A safe operating space for humanity. Nature 461:472–475. https://doi.org/10.1038/461472a

Rockström JW et al (2009b) Planetary boundaries: exploring the safe operating space for humanity. Ecol Soc 14(2):32 [online]. URL: http://www.ecologyandsociety.org/vol14/iss2/art32/

Soares-Filho B et al (2014) Cracking Brazil's forest code. Science 344(6182):363–364. Cited on http://www.sciencemag.org/content/344/6182/363.short. Accessed: 17 Jan 2021

Steffen W et al (2015) Planetary boundaries: guiding human development on a changing planet. Ciences 347:11. https://www.science.org/doi/10.1126/science.1259855

Stockholm Resilience Centre (SRC) (2022) Planetary boundaries, 17 Feb 2022. Cited on https://www.stockholmresilience.org/research/planetary-boundaries.html

UNEP (2012) Scenarios and sustainability transformation. In: Global environment outlook, vol 5. Progress Press, Valleta, Malta, 528p

UNESCO (2021) Gobernanza del agua en América Latina y el Caribe. Cited on https://es.unesco.org/fieldoffice/montevideo/agua/fasesPHI. Acessado em 07 Sept 2021

United Nations Convention to Combat Desertification (UNCCD) (2020) The Great Green Wall implementation status and way ahead to 2030. Bonn, Germany, 45p

Printed in the United States
by Baker & Taylor Publisher Services

Printed in the United States
by Baker & Taylor Publisher Services